NATIONAL PORT STRATEGY ASSESSMENT: Reducing Air Pollution and Greenhouse Gases at U.S. Ports

Appendices

List of Appendices to National Assessment of Port Strategies: Reducing Air Pollution and Greenhouse Gases at U.S. Ports

Appendix A.   Baseline Emission Inventory Methodology

A.1. Introduction

This assessment included the development of representative, national scale inventories for the baseline and Business as Usual (BAU) cases for different pollutants and years, followed by the analysis of various strategies to reduce port-related mobile source emissions. This appendix details the methodology used to develop the baseline emission inventories for the calendar year 2011.

Separate inventories for various pollutants were developed for the drayage trucks, rail, cargo handling equipment (CHE), harbor craft, and ocean going vessels (OGV) sectors. The following pollutants were included in these inventories: nitrogen oxides (NOx), fine particulate matter (PM$_{2.5}$), volatile organic compounds (VOCs), sulfur dioxide (SO$_2$), carbon dioxide (CO$_2$), black carbon (BC), acetaldehyde, benzene, and formaldehyde. Note that the selected air toxics (acetaldehyde, benzene, and formaldehyde) were only analyzed for the non-OGV sectors and SO$_2$ was only analyzed for the OGV sector. In general, inventories were developed for each port analyzed in this assessment using national scale methodology and data, although port-specific data and adjustments were included where available and are noted where appropriate in this appendix. This assessment does not provide specific data for local decision-making at individual ports or specific neighborhoods.

EPA developed this national scale assessment based on estimated emissions from a representative sample of seaports, which are listed in Table A-1. Several of the ports have publicly available emission inventories that were used to improve the information applied in this assessment. Ports were chosen to represent typical deep sea ports in the United States, so it does not include any inland freshwater ports. However, it is expected that emission reduction strategies would be applicable to other seaports, Great Lakes and inland river ports, or other freight and passenger facilities with similar mobile source profiles.

It should be noted that this project is not intended to provide port-specific results. EPA did not consult with the 19 ports before port areas were selected for this assessment. The results from the 19 ports were combined throughout this assessment to present a national scale picture of how emissions change between the baseline, future BAU case, and what can be done to further reduce emissions.

Table A-1. Ports Selected for Assessment

No.	Port	Published Inventory Available
1	Port of New York and New Jersey	Yes
2	Port of New Orleans	No
3	Port of Miami	No
4	Port of South Louisiana	No
5	Port of Seattle	Yes
6	Port of Baton Rouge	No
7	Port Arthur	Yes
8	Port of Portland, OR	Yes
9	Port of Mobile	No
10	Port of Houston	Yes
11	Port of Baltimore	No*
12	Port of Hampton Roads (Norfolk)	Yes
13	Port of Philadelphia	Yes
14	Port of Charleston	Yes
15	Port of Corpus Christi	Yes
16	Port Tampa Bay	No
17	Port of Savannah	No
18	Port of Coos Bay, OR	No
19	Port of San Juan, PR	No

* The Port of Baltimore published a CHE inventory, but it was not available during the timeframe for inclusion in this assessment.

While deciding which ports to incorporate in this national scale assessment, EPA selected a representative sample of ports that was intended to be diverse. EPA considered several factors in selecting this sample of port areas: the geographic location of a given seaport; the type and size of different ports; the availability of port emission inventories; and whether or not a port was located in or adjacent to a nonattainment or maintenance area for the national ambient air quality standards (NAAQS). The geographic distribution of the ports selected for assessment is shown in Figure A-1.[1] The assessment was also based on other publically available activity data for certain port areas that have been used for other previous EPA analyses, as noted in throughout this appendix and Appendix B.

---

[1] EPA notes that the Port of Los Angeles and the Port of Long Beach were initially selected for inclusion in this national scale assessment. However, EPA decided to not include emission estimates for these port areas to avoid biasing the national scale results. Many of the strategies in this report are already being implemented at these ports, and there are also port-specific baseline and future emission inventories in place, all of which could potentially impact the final results. EPA consulted on this decision with the MSTRS Ports Workgroup, which included several ports, government agencies, community groups, and other policy and technical experts.

Figure A-1. Geographic Distribution of Selected Ports[2]

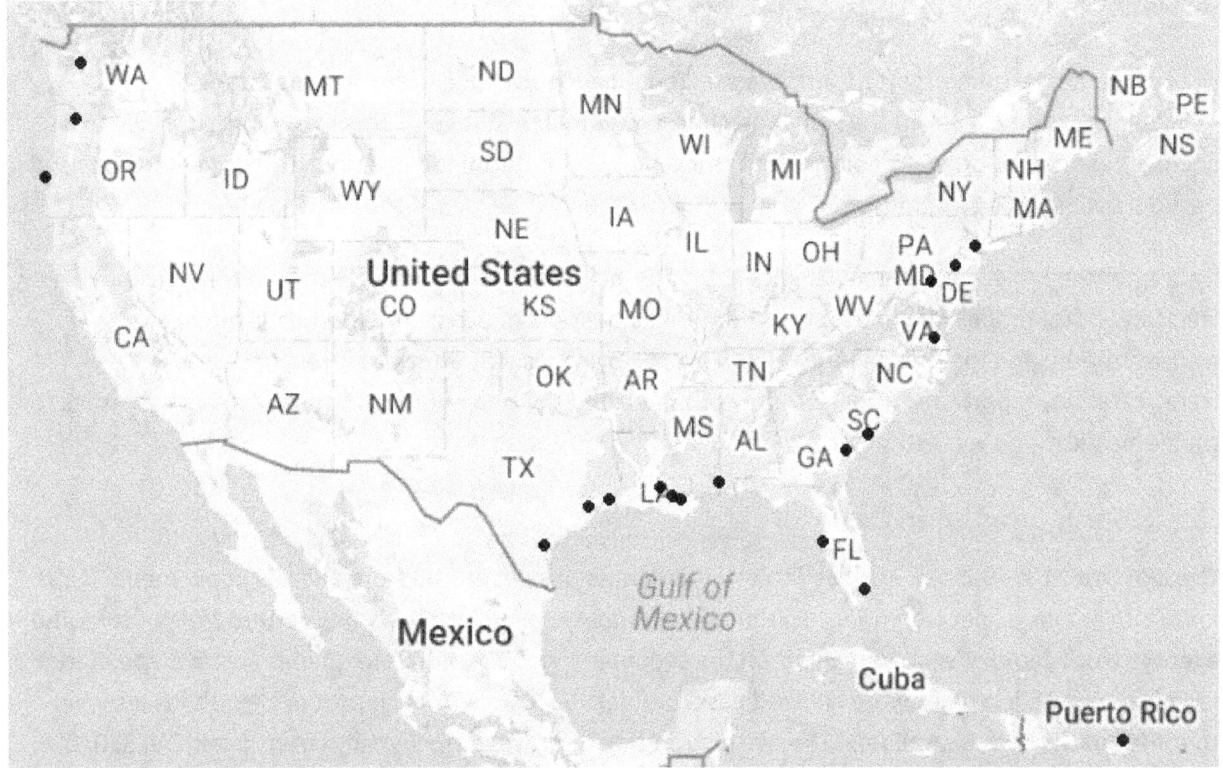

The following sections detail how the baseline inventories were developed for each sector.

A.2. Drayage Trucks

The 2011 base year drayage emissions were modeled using EPA's DrayFLEET Model. The truck activity is based on the total tonnage and TEUs moved at a port allocated by mode to drayage using FHWA's Freight Analysis Framework (FAF).

A.2.1. Data Sources

Two primary types of activity data were collected for the analysis: tonnage[3] and twenty foot equivalent units[4] (TEUs) by port. Activity data from the US Army Corps of Engineers (USACE) Waterborne Commerce Statistics on TEUs and tonnage by port were collected for the base year 2011. For containerized movements, data on the number of TEUs were used to estimate truck drayage activity. TEUs were translated into truck movements based on estimates of TEUs per container. The USACE Waterborne Commerce Statistics includes data on domestic empty containers, but not foreign empty

---

[2] Map data: Google, INEGI.

[3] Available at: http://www.navigationdatacenter.us/wcsc/by_portnames11.html.

[4] Available at: http://www.navigationdatacenter.us/db/wcsc/archive/xls/man11/.

containers. Data on foreign empty containers were collected separately from ports or other industry sources.[5]

For non-containerized cargo, the analysis used data on the tonnage of freight originating and terminating in the port from the ACE Waterborne Commerce Statistics. This cargo was classified as bulk, liquid, container, or other. The number of truckloads was determined based on the cargo densities and payload estimates by commodity.

Data from the FAF for 2012 were used to estimate the percentage of containers and tonnage moving by truck and other modes for each port. FAF identifies the port of export, the domestic mode of transportation, and the foreign mode. For freight moving via water for the foreign mode of transportation, exports and imports were combined and the percentage of tonnage moving by truck for the domestic mode was estimated.

A.2.2. Modeling Approach

The EPA DrayFLEET Model[6] was used to estimate emissions generated from all truck movements. TEUs and TEUs per container were input on the primary inputs page. For non-containerized freight, tons and average truck cargo weight were input on the secondary inputs page. Gate queues and average marine terminal transaction times were entered based on data available for each port or default values when this information was unavailable. The age distribution of drayage trucks came from the MOVES2010b national default age distribution for combination short-haul trucks.

A 0.5 km boundary was included in this analysis. Drayage emissions for the port and within a 0.5 km buffer outside the port were modeled separately. This was accomplished by estimating the distance drayage vehicles travel inside the port and the distance they travel outside the port within a 0.5 km of the port boundary. A visual inspection of port maps was made to estimate these distances. In the case of emissions outside of the port, a distance longer than 0.5 km was selected for some ports to account for route circuitry around the port boundary.

One model scenario was set up to include the total drayage distance, while a second scenario included only the distance within the port. The difference in emissions estimates between these models was used to estimate emissions outside of the port boundaries but within the 0.5 km buffer.

The DrayFLEET Model estimates emissions for $PM_{2.5}$, NOx, HC, CO, and CO_2. Emissions for the air toxics formaldehyde, benzene, and acetaldehyde were estimated as a fraction of HC emissions based on diesel speciation profiles calculated from running MOVES2010b, as shown in Table A-2. These fractions vary depending on whether or not VOC is controlled (model year 2007 and later). Weighted speciation

---

[5] Available at: http://aapa.files.cms-plus.com/Statistics/NORTH%20AMERICAN%20PORT%20CONTAINER%20TRAFFIC%202011.pdf.

[6] Available at: http://www.epa.gov/smartway/forpartners/documents/drayage/420b12065.pdf.

factors were calculated for 2011 based on the percent of 2007 model year and greater trucks in the fleet. Approximately 86% of drayage trucks in 2011 were model year 2006 or earlier.

Table A-2. Diesel Truck Air Toxic Speciation Profiles Based on MOVES2010b[7]

Pollutant	Toxic/VOC without Control	Toxic/VOC with Control
Acetaldehyde	0.035559	0.06934
Benzene	0.007835	0.01291
Formaldehyde	0.078225	0.21744

Black carbon (BC) emissions were estimated as a 77% of $PM_{2.5}$ emissions, consistent with EPA's Report to Congress.[8]

A.3. Rail

The 2011 baseline inventory of rail emissions is based on two primary sources: published port emission inventories and the 2011 National Emission Inventory (NEI). The inventory for all 19 ports considered here is drawn from one of these two sources, as described below.

A.3.1. Data Sources

A.3.1.1. Published Port Inventory Data

EPA performed a review of available port inventories and determined that for four ports, the on-terminal estimates from published rail emission estimates are a better match for this project's scope than values derived from the NEI. Table A-3 lists these four ports and the information included in this analysis from each inventory.

---

[7] U.S. Environmental Protection Agency, *MOVES2010b: Additional Toxics Added to MOVES.* EPA-420-B-12-029a, May 2012, Sec 3.1.1. Available at: http://www.epa.gov/otaq/models/moves/documents/420b12029a.pdf.

[8] U.S. Environmental Protection Agency, *Report to Congress on Black Carbon*, EPA-450/R-12-001, March 2012, p. 87.

Table A-3. Rail Emissions from Published Port Inventories Included in Baseline Inventory

Port	Published Inventory	Rail Type Included
Charleston	South Carolina State Ports Authority 2011 Emissions Inventory Update (April 2013)	"On-Terminal" results for Line-Haul and Switcher Locomotives
Norfolk	Port of Virginia 2011 Comprehensive Air Emissions Inventory Update (January 2013)	"On-Terminal" results for Line-Haul and Switcher Locomotives
Seattle	Puget Sound Maritime Air Emissions Inventory, August 2012 (May 2013)	Port of Seattle estimates of switcher and Line-Haul Locomotives, with the understanding that the line-haul inventory only includes emissions from "near port/adjacent rail yards."
New York / New Jersey	Port of New York and New Jersey Port Commerce Department 2012 Multi-Facility Emissions Inventory (August 2014)	Switcher emissions only.

A.3.1.2. NEI Data

For the 15 ports for which published inventories were not used, the methodology for estimating the baseline rail emissions relied on results from EPA's 2011 v1 NEI.[9] In the NEI, emissions are reported by source classification codes (SCCs). Table A-4 shows those related to the rail sector (line-haul class I–III and yard locomotives) included in this analysis and the category in which each is classified in the NEI. Yard locomotives are categorized in the NEI as point sources.[10] It should be noted that the SCCs do not distinguish between the types of rail activities; therefore, there is no way to explicitly differentiate the port-related locomotive emissions from other rail emissions in the NEI.

Table A-4. Relevant SCCs for Port Rail Inventory Analysis

SCC	NEI Data Category	Description
2285002006	Nonpoint	Line-haul Locomotives: Class I Operations
2285002007	Nonpoint	Line-haul Locomotives: Class II / III Operations
28500201	Point	Yard Locomotives

The 2008 NEI rail emission inventory was based on a methodology developed by the Eastern Regional Technical Advisory Committee (ERTAC), a group coordinating the efforts of 27 eastern state air quality agencies. The approach allocates locomotive emissions into three categories: Class I line-haul, Class II and III line-haul, and Class I switchers at rail yards.[11] The emission totals were based on activity levels

---

[9] Available at: https://www.epa.gov/air-emissions-inventories/2011-national-emissions-inventory-nei-data.

[10] There is an additional SCC in the NEI for yard locomotives (2285002010: Nonpoint Yard Locomotives). However, all EPA estimates in the NEI for yard locomotive emissions are recorded as point sources (SCC 28500201).

[11] Rail classes are defined by the Surface Transportation Board. Class I refers to rail companies with operating revenues of $433.2 million or more in 2011. Class II typically refers to what is more commonly known as regional railroads. Class III are short line railroads that generate less than $20 million in revenue (1991 dollars). The ERTAC methodology excludes Amtrak activity because of the difference in activity characteristics from the other rail classes.

either reported publicly to federal agencies or provided confidentially to ERTAC by the rail companies. ERTAC also obtained proprietary fleet mix data from the Class I railroads to weight emission factors for rail segments on which those companies operated trains. The 2011 NEI rail emissions were estimated from the 2008 inventory using an annual growth rate of -2.475% for Class I railroads and -8.37% for Class II and III railroads.[12]

NEI rail yard emissions and locations are based on data from the Federal Railroad Administration and an analysis of the Bureau of Transportation Statistics public rail network.[13] NEI shapefiles for rail activity include rail line geometries, identifier codes, county FIPS codes, lengths, track rights, and other geographic and administrative attributes. Shapefiles for ports and rail lines were obtained from the NEI website.[14] Updates to these shapefiles were made using the latest available data for Houston, Miami, Norfolk, Savannah, and New York/New Jersey.

A.3.2. Modeling Approach

For ports that had published inventories, those data were included in the aggregate inventories reported in this assessment as described in Table A-3. However, in some cases these published inventories did not include all the pollutants needed for this assessment (e.g., VOC, carbon dioxide, black carbon, acetaldehyde, benzene, and formaldehyde). For example, the Charleston and Norfolk inventories do not include VOC emissions; instead, these emissions were calculated from reported total hydrocarbon values using EPA conversion factors.[15] Seattle's CO_2 emissions were assumed to be 99% of the reported CO_2e emissions. Charleston reports neither CO_2 nor CO_2e. In this case, CO_2 emissions are estimated from CO emissions by regressing the two pollutants across all other reported inventories. This is justified by the high correlation ($r^2=0.984$) and the fact that most engines do not include aftertreatment for CO emissions that would substantially alter this ratio.

None of the inventories included black carbon (BC) or any of the three air toxics, acetaldehyde, benzene, and formaldehyde. In all cases, black carbon locomotive emissions were estimated to be 77% of $PM_{2.5}$ values, as reported in EPA's Report to Congress.[16] Air toxic emissions were estimated VOC inventories using speciation factors, which were calculated based on the NEI 2011 v1 total switcher or line-haul VOC and air toxics emissions from the 19 ports. Table A-5 shows the resulting speciation factors.

---

[12] Methodology for updating 2008 NEI locomotive inventory for 2011v1 NEI is described in ERG Memorandum to EPA "Development of 2011 Railroad Component for National Emissions Inventory," September 5, 2012. Available at: ftp://ftp.epa.gov/EmisInventory/2011/doc/2011nei_Locomotive.pdf.

[13] A description of the methodology used to assess and allocate rail yard emissions in the NEI can be found in EPA (May 2011) report "Documentation for Locomotive Component of the National Emissions Inventory Methodology" prepared by Eastern Research Group. Available for download at ftp://ftp.epa.gov/EmisInventory/2011/doc/2008nei_locomotive_report.pdf.

[14] Available at: http://www.epa.gov/ttn/chief/net/2011inventory.html.

[15] U.S. Environmental Protection Agency, *Conversion Factors for Hydrocarbon Emission Components*, NR-002d, July 2010.

[16] U.S. Environmental Protection Agency, *Report to Congress on Black Carbon*, EPA-450/R-12-001, March 2012, p. 87.

Table A-5. Calculated Air Toxic Speciation Factors for Rail

BAU Source	VOC	Formaldehyde	Benzene	Acetaldehyde
Rail Line	1	0.0416	0.00249	0.0181
Rail Yard	1	0.0260	0.00155	0.0113

For ports that relied on the NEI, additional calculations were required as the NEI reports locomotive emissions only down to the county level. The steps to calculate locomotive emissions associated with rail lines and rail yards in and around each specific port are described below.

Using the port and rail shapefiles, the rail lines were mapped in a GIS program. The rail yards were also mapped using latitude and longitude data associated with each rail yard in the NEI point source database. Rail lines and rail yards located within the study area were selected. Rail lines that cross the study area boundary were "clipped" in the GIS program so that only the portions of the rail segments that fall within the study area were selected. The GIS program was also used to measure the lengths of each rail line segment in the study area.[17] Attribute tables associated with each rail line segment and rail yard within the study area were exported for analysis.

The exported attribute data include unique identifiers for each rail line (*FRAARCID*) and rail yard (*eis_facili*). The rail line attributes also include data on the length of each segment; for segments that were clipped to the study area boundary, the GIS program recalculates the new length of the rail line.

In the 2011 NEI Documentation section of the NEI website, a table with the fractions of county emissions for rail is available for download. This table includes a field with each rail segment's unique identification code (*FRAARCID*), the SCCs for line-haul locomotives (see Table A-4) and the fraction of the total county locomotive emissions allocated to each rail segment by SCC. Using the rail segment unique identifier, the fractions of county emissions were linked to the matching rail segment.[18] For segments that were clipped at the study boundary, the ratio of the clipped length to the original length was used to adjust the emissions fraction proportionally; for example, a rail line with only half of its length within the study boundary had its associated fraction of county emissions halved.

The 2011 NEI Nonpoint database identifies counties in the United States using Federal Information Processing Standard (FIPS) codes. Counties are also identified in the rail shapefile attribute tables by FIPS codes. Using these codes, the total county locomotive emissions were joined to each rail segment. The county emissions were then multiplied by the fractions (or clipped fractions) identified earlier to calculate the portion of county emissions allocated to each rail segment.

---

[17] Note that the attribute data already in the rail line shapefile include lengths for each rail segment (as per *MILES*). However, for this analysis, the segment lengths were re-measured in the GIS program to ensure consistency with the port shapefile used to establish the study area.

[18] Note that not all rail segments are listed in the county fraction table; thus, in this inventory, they have no emissions associated with them.

The 2011 NEI Point database includes emission levels (*total_emissions*) by pollutant (*pollutant_cd*) for each point source facility (*eis_facility_site_id*). The emissions for each rail yard located within the study area were looked up using the facility identification codes (*eis_facility_site_id* in the NEI Point dataset and *eis_facili* in the rail shapefile attribute table).

Using these steps, the emissions for the rail lines and rail yards in the study area were estimated. However, these emissions were limited to those pollutants included in the 2011 NEI version 1, which includes NOx, VOCs, $PM_{2.5}$, benzene, acetaldehyde, and formaldehyde.[19] The other pollutants assessed in this project but not included in the NEI (black carbon and carbon dioxide) were calculated separately.

Consistent with EPA's Report to Congress, black carbon emissions are assumed to be 77% of $PM_{2.5}$ emissions.[20]

CO_2 emissions must be estimated from the other species available in the 2011 NEI. Emissions factors from the 2009 EPA report "Current Methodologies in Preparing Mobile Source Port-Related Emission Inventories" were applied as described below.[21] Table A-6 shows the locomotive emission factors from the report.

Table A-6. Line-haul Locomotive Emission Factors (g/bhp-hr)

Calendar Year	HC	CO	NOx	PM_{10}	CO_2	N_2O	CH_4
2005	0.48	1.28	13	0.32	483	0.04	0.013
2006	0.47	1.28	12.79	0.32	483	0.04	0.013
2007	0.45	1.28	12.15	0.30	483	0.04	0.013
2008	0.42	1.28	11.14	0.28	483	0.04	0.013
2009	0.39	1.28	10.17	0.26	483	0.04	0.013

Based on the consistent values of the CO and CO_2 emission factors over the five years backcasted in the report, CO was used as a basis for scaling emissions for CO_2, using the ratio between the two factors. The CO_2 emissions factor is 377 times greater than the CO emission factors (483/1.28). Thus, to calculate carbon dioxide emissions, carbon monoxide values for each rail line and rail yard were multiplied by 377.

A.4. Cargo Handling Equipment

For the cargo handling equipment sector, existing information from the NEI or other models could not be used to estimate the 2011 baseline inventory. Both the NEI and the NONROAD model allocate emissions to counties but do not allocate emissions to localized areas like ports well. A regression

---

[19] This work is based on v1 of the 2011 NEI, the most current published version of the dataset at the time of analysis.

[20] U.S. Environmental Protection Agency, *Report to Congress on Black Carbon*, EPA-450/R-12-001, March 2012, p. 87.

[21] U.S. Environmental Protection Agency, *Current Methodologies in Preparing Mobile Source Port-Related Emission Inventories*, April 2009.

analysis was performed using published port emission inventories to predict emissions at all 19 ports of interest. This methodology is designed to avoid issues with direct calculation models and NEI values and instead extrapolate known CHE inventories to ports with unknown values.

A.4.1. Data Sources

A.4.1.1. NEI

Using the 2011 NEI emissions estimates was not feasible for CHE. NEI values are reported at the county level and by source classification code (SCC). SCC codes do not match well to the types of equipment moving goods at ports, and the SCCs corresponding to equipment at ports may also be in use within the surrounding county for applications not associated with port activity. Further, while SMOKE[22] has allocation surrogates for these SCCs, these surrogates do not account for activity at ports. Therefore, countywide values from the published version of EPA's 2011 v1 NEI[23] were instead used as a quality assurance check on the results.

A.4.1.2. Published Port Emission Inventories

Table A-7 lists those ports with sufficiently recent and comprehensive CHE emission inventories to use in this analysis. All are for calendar year 2011, except Oakland, which is 2012. Port Arthur (2000), Portland (2000), Corpus Christi (1999), Philadelphia (2003), and Houston (2007) also have published inventories, but were considered too old to use here. New York/New Jersey (2008) also has a published CHE inventory, but it is both old and only covers container terminals, which account for only about 16% of goods handled by weight and was considered this too limited to use for regression. Richmond, CA, also has a published inventory (2005), but it is both too old and limited to use in this regression analysis.

Table A-7. Published Port Inventories Used in This CHE Emission Regression Analysis

Port		
Anacortes	Tacoma	Port of Virginia
Everett	Port of Los Angeles	Charleston
Olympia	Port of Long Beach	Oakland
Seattle		

A.4.1.3. US Army Corps of Engineers Waterborne Commerce

Data for the 2011 baseline year came from Waterborne Commerce of the United States (WCUS), collected and published by the US Army Corps of Engineers. These data included cargo throughput, in terms of tonnage[24] and TEUs[25] and includes both domestic and international movements.[26] USACE TEU data were only

---

[22] SMOKE is the Sparse Matrix Operator Kernel Emissions Model. It is an emission inventory processing system used in certain photochemical modeling applications, such as the Community Multi-Scale Air Quality Model (CMAQ).

[23] Available at: https://www.epa.gov/air-emissions-inventories/2011-national-emissions-inventory-nei-data.

[24] Available at: http://www.navigationdatacenter.us/db/wcsc/archive/xls/man11/.

[25] Available at: http://www.navigationdatacenter.us/wcsc/by_portnames11.html .

[26] Available at: http://www.navigationdatacenter.us/data/datawcus.htm and http://www.navigationdatacenter.us/data/datappor.htm.

available for the 84 ports USACE considers "principal." Cargo throughput for most domestic waterways is available through the WCUS dataset. These data track cargo throughput (in thousands of tons) in 146 commodity categories. For this analysis, these commodity categories were grouped into four conveyance methods: dry and break bulk, liquid bulk, containerized (TEUs), and other (principally roll-on/roll-off and automobile). Passenger is not considered a "conveyance method" and not included here. Table A-8 lists the USACE commodity types tracked in the WCUS dataset and the four conveyance methods used.

Table A-8. Commodity Types and Matching Conveyance Methods

Conveyance Method	Product
Bulk	Aluminum, Aluminum Ore, Animal Feed, Prep., Barley & Rye, Building Stone, Cement & Concrete, Clay & Refrac. Mat., Coal Coke, Coal Lignite, Cocoa Beans, Copper, Copper Ore, Corn, Dredged Material, Ferro Alloys, Fert. & Mixes NEC, Flaxseed, Forest Products NEC, Fuel Wood, Grain Mill Products, Gypsum, Hay & Fodder, I&S Bars & Shapes, I&S Pipe & Tube, I&S Plates & Sheets, I&S Primary Forms, Iron & Steel Scrap, Iron Ore, Lime, Limestone, Lumber, Machinery (Not Elec), Manganese Ore, Marine Shells, Metallic Salts, Misc. Mineral Prod., Molasses, Non-Ferrous Ores NEC, Non-Ferrous Scrap, Non-Metal. Min. NEC, Non-Metal. Min. NEC, Oats, Ordnance & Access., Peanuts, Petroleum Coke, Phosphate Rock, Pig Iron, Primary I&S NEC, Primary Wood Prod., Pulp & Waste Paper, Radioactive Material, Sand & Gravel, Slag, Soil & Fill Dirt, Sorghum Grains, Starches, Gluten, Glue, Sugar, Sulphur, (Dry), Unknown or NEC, Waste and Scrap NEC, Waterway Improv. Mat, Wheat, Wheat Flour, Wood Chips, Wood in the Rough
Container	Alcoholic Beverages, Animals & Prod. NEC, Bananas & Plantains, Coffee, Coloring Mat. NEC, Cotton, Dairy Products, Electrical Machinery, Empty Containers, Fab. Metal Products, Farm Products NEC, Food Products NEC, Fruit & Nuts NEC, Glass & Glass Prod., Groceries, Inorg. Elem., Oxides, & Halogen S, Manufac. Prod. NEC, Manufac. Wood Prod., Meat, Fresh, Frozen, Meat, Prepared, Medicines, Natural Fibers NEC, Newsprint, Nitrogen Func. Comp., Paper & Paperboard, Paper Products NEC, Perfumes & Cleansers, Pesticides, Pigments & Paints, Plastics, Rice, Rubber & Gums, Rubber & Plastic Pr., Smelted Prod. NEC, Soybeans, Textile Products, Tobacco & Products, Vegetables & Prod.
Liquid	Acyclic Hydrocarbons, Alcohols, Ammonia, Asphalt, Tar & Pitch, Benzene & Toluene, Carboxylic Acids, Chem. Products NEC, Chemical Additives, Crude Petroleum, Distillate Fuel Oil, Fruit Juices, Gasoline, Inorganic Chem. NEC, Kerosene, Liquid Natural Gas, Lube Oil & Greases, Naphtha & Solvents, Nitrogenous Fert., Oilseeds NEC, Organic Comp. NEC, Organo-Inorganic Comp., Other Hydrocarbons, Petro. Jelly & Waxes, Petro. Products NEC, Phosphatic Fert., Potassic Fert., Residual Fuel Oil, Sodium Hydroxide, Sulphur (Liquid), Sulphuric Acid, Tallow, Animal Oils, Vegetable Oils, Water & Ice, Wood & Resin Chem.
Other	Aircraft & Parts, Explosives, Fish (Not Shellfish), Fish, Prepared, Shellfish, Ships & Boats, Vehicles, Vehicles & Parts

A.4.2. Modeling Approach

A regression model was developed based on the observed relationship between port cargo throughput and CHE emissions. The recent CHE emission inventories for the ports listed in Table A-7 were collected, and gaps that existed for certain pollutants in some of the inventories were filled as described in the next section. Several different regression models were explored to determine correlations between NO_x, VOC, $PM_{2.5}$, and CO_2 emissions in tons per year against cargo throughput:

- Method 1: regressed total CHE emissions for each pollutant against the tonnage in the four conveyance types

- Method 2: regressed total CHE emissions for each pollutant against the total non-container tonnage and number of TEUs

- Method 3: regressed total CHE emissions for each pollutant against total tonnage throughput only, excluding conveyance type

- Method 4: represents an unweighted average of the predictions from the above three methods

Method 4 was used in this analysis as the best option based on engineering judgement. This regression model has similar limitations as with other sectors that rely on NEI values. As with those sectors, it does not allow characterization of emissions by equipment age, fuel type, terminal type, existing use of control technology, or other discriminators. In this analysis, all regressions were performed against total CHE emissions rather than against individual equipment types due to data limitations.[27]

In cases where different definitions of hydrocarbons were included in the published inventories, all species were converted to VOC using NONROAD factors.[28] For published inventories where only CO_2e was included, CO_2 was estimated as 99% of CO_2e. BC emissions were taken as 77% of the regressed $PM_{2.5}$ emissions, consistent with EPA's Report to Congress.[29]

For all modeled ports, VOC, $PM_{2.5}$, NOx, and CO_2 were determined from the regression model. Benzene, acetaldehyde, formaldehyde, and BC emissions were then determined from the VOC or $PM_{2.5}$, as appropriate, either from the regressed estimates (for modeled ports) or from the published values (where available). Speciation factors for benzene, acetaldehyde, and formaldehyde relative to VOC were derived from values in version 1 of the 2011 NEI. National total emissions of the air toxics and VOCs were summed for the relevant CHE SCCs (see Table A-9) and ratios were calculated for each. These speciation factors are very similar to the national averages considered for each SCC individually, which showed little variation, but accommodated the different fuel types in use in CHE nationally.

Four calculated port inventories were compared against countywide NEI values for certain SCCs to confirm the results as reasonable. Only the emissions from SCCs that potentially may operate on ports were included in this comparison. See Table A-9 for a complete listing of SCCs included. The calculated inventories for the four ports were compared to the countywide NEI values for the associated areas as listed in Table A-10.

---

[27] Note that this regression includes Los Angeles and Long Beach as inputs. These ports were later removed from the assessment's results but not until after this analysis had been conducted. It is possible that the CHE emissions profile at these ports differs from other, non-California ports and may influence the results.

[28] NONROAD HC Conversion Factors. Available at http://www.epa.gov/otaq/models/nonrdmdl/nr-002.pdf.

[29] U.S. Environmental Protection Agency, *Report to Congress on Black Carbon*, EPA-450/R-12-001, March 2012, p. 87.

Table A-9. 2011 NEI CHE Types by SCC

SCC	SCC Level Four	SCC Level Three
2270002015	Rollers	Construction and Mining Equipment
2270002027	Signal Boards/Light Plants	Construction and Mining Equipment
2270002036	Excavators	Construction and Mining Equipment
2270002045	Cranes	Construction and Mining Equipment
2270002051	Off-highway Trucks	Construction and Mining Equipment
2270002060	Rubber Tire Loaders	Construction and Mining Equipment
2270002063	Rubber Tire Tractor/Dozers	Construction and Mining Equipment
2270002066	Tractors/Loaders/Backhoes	Construction and Mining Equipment
2270002069	Crawler Tractor/Dozers	Construction and Mining Equipment
2270002072	Skid Steer Loaders	Construction and Mining Equipment
2270002075	Off-highway Tractors	Construction and Mining Equipment
2270003010	Aerial Lifts	Industrial Equipment
2270003020	Forklifts	Industrial Equipment
2270003030	Sweepers/Scrubbers	Industrial Equipment
2270003050	Other Material Handling Equipment	Industrial Equipment
2270003070	Terminal Tractors	Industrial Equipment
2270006005	Generator Sets	Commercial Equipment
2270006010	Pumps	Commercial Equipment
2270006015	Air Compressors	Commercial Equipment
2270006025	Welders	Commercial Equipment

Table A-10. NEI Areas Used in Regression Comparison

Port	NEI Areas of Comparison
Port of Houston	Harris County, TX
Port of New Orleans	St. Charles, Jefferson, and Orleans Parishes
Port of Portland	Multnomah County, OR
Port of Savannah	Chatham County, GA

The comparison showed that the calculated port inventories were between 2% and 54% of the corresponding NEI inventories for CO_2, NOx, $PM_{2.5}$, and VOCs, with percentages varying by port and pollutant. For the air toxics studied, three of the port inventories were between 3% and 32% of their corresponding NEI inventories, but one port was calculated to be approximately 185% of its NEI inventory for air toxics. However, without additional information on other construction, mining, industrial, and commercial activity present in each of these areas, it is difficult to comprehensively assess the reasonableness of these values. Based on engineering judgement, the regression model was determined to be adequate for the purposes of this assessment.

A.5. Harbor Craft

This section discusses the methodology used to develop the 2011 baseline harbor craft sector emission inventory. The term "harbor craft" is used synonymously for all Category 1–Category 2 vessels (C1/C2). For this sector, all emissions are determined based on the 2011 NEI. Existing port inventories were not used to assess harbor craft emissions because most port inventories only included harbor craft emissions related to their own port operations, and did not include, for example, harbor craft activity that could be related to private terminals.

A.5.1. Data Sources

The primary data source for this sector was a draft of 2011 NEI v2. In the NEI, emissions are reported by source classification codes (SCCs). Table A-11 shows those related to the harbor craft emission sectors included in this analysis.

Table A-11. Relevant SCCs for Port Inventory Analysis

SCC	Emission Type Code*	Description
2280002100	M	Harbor Craft (C1/C2) at Port
2280002200	C	Harbor Craft (C1/C2) underway

* Emission type codes for C1/C2 vessels are defined as M for maneuvering (in port) and C for cruise (out of port).

The NEI contains some measure of both state, local, and tribal (SLT) estimates and EPA estimates of emissions. Consistent with the national focus of this analysis, and to insure that a consistent methodology was used for each area, only the EPA derived emission estimates were used instead of the SLT submitted estimates.

A.5.2. Modeling Approach

The NEI includes an allocation of marine vessel emissions to GIS shapefiles for near-port operating modes (maneuvering) and out-of-port (cruise) modes for harbor craft. These NEI-defined polygons allocate harbor craft emissions to either port-encompassing polygons[30] or open water shipping lane polygons,[31] respectively. The polygons of port boundaries are based on maps provided directly from ports, from local port authorities and port districts, and from satellite imagery and GIS tools. Polygons were created on land, bordering waterways, and coastal areas, and were split by county boundary such that no shape file crosses county lines and county total emission can be easily summed.[32] These are referred to in this report as the "NEI shapes."

This analysis identified all NEI shipping lane shapes within 5 km of each port facility,[33] where 5 km represents the outermost edge of cruising activity for the purposes of this assessment. All NEI shipping lane shapes were clipped at this 5 km buffer edge and the portion of each shape within the buffer was recorded. There were two exceptions to this exercise for the Port of Miami and Port of Hampton Roads.

---

[30] Available at: http://www.epa.gov/ttn/chief/eis/2011nei/2011_ports_shapefile.zip.

[31] Available at: http://www.epa.gov/ttn/chief/eis/2011nei/shippinglanes_112812_shapefile.zip.

[32] U.S. EPA, *2011 NEI Technical Support Document*, November 2013, available at: https://www.epa.gov/air-emissions-inventories/2011-nei-technical-support-document.

[33] In cases where ORD shapes are taken representing the port, the 5 km buffer for shipping activity extends from the boundary of these shapes, not the original NEI shapes.

For the Port of Miami, the cruising activity in the Port of Miami River was not included. Additionally, NEI shapes for Newport News were excluded from the Port of Hampton Roads.

To calculate the harbor craft inventory, at port maneuvering emissions for NOx, $PM_{2.5}$, VOC, and CO reported in the draft 2011 v2 NEI were summed by harbor craft type for each port location. Cruising emissions for the same pollutants were determined by summing the NEI emissions for each port's associated shipping lane shapes. In cases where a shipping lane shape extends beyond the 5 km buffer, a proportional fraction of the shape's emissions were attributed to the port. For example, if 10% of a port's shipping lane shape lies within the 5 km buffer, 10% of the total cruising mode emissions associated with that shape are included in the inventory for that port.

The draft version 2 of the 2011 NEI that was used for this analysis did not contain estimates for CO_2, air toxics, or black carbon. However, a previous study[34] indicated that CO_2 and VOC show a direct correlation, more so than for CO or other pollutant. Therefore, a scaling factor for VOC to CO_2 emissions was calculated based on a previous iteration of the draft 2011 version 2 NEI. The scaling factor was determined to be 3,247.9 tons CO_2 per ton VOC, which did not vary by vessel type.

Similarly to CO_2, the air toxics were speciated from VOC using factors calculated from a previous iteration of the draft 2011 version 2 NEI. Table A-12 shows these speciation factors.

Table A-12. Select Air Toxic Speciation Factors from VOC Emissions for Harbor Craft

SCC	Pollutant	Speciation Factor
2280002100	Acetaldehyde	0.0557235
2280002200	Acetaldehyde	0.046436
2280002100	Benzene	0.015258
2280002200	Benzene	0.012715
2280002100	Formaldehyde	0.1122
2280002200	Formaldehyde	0.0935

Consistent with EPA's Report to Congress, black carbon emissions are assumed to be 77% of $PM_{2.5}$ emissions.[35]

A.6. Ocean Going Vessels

This section details the methodology used for developing the ocean going vessel (OGV) 2011 baseline emission inventories. It is based primarily upon the methodology used for the Category 3 Marine Engine Rulemaking[36] (C3 RIA). Using the C3 RIA modeling approach, the OGV emission inventories were

---

[34] U.S. Environmental Protection Agency, *Current Methodologies in Preparing Mobile Source Port-Related Emission Inventories*, April 2009.

[35] U.S. Environmental Protection Agency, *Report to Congress on Black Carbon*, EPA-450/R-12-001, March 2012, p. 87.

[36] U.S. Environmental Protection Agency, *Regulatory Impact Analysis: Control of Emissions of Air Pollution from Category 3 Marine Diesel Engines,* EPA Report EPA-420-R-09-019, December 2009. Available at: http://www.epa.gov/otaq/regs/nonroad/marine/ci/420r09019.pdf.

calculated using energy-based emission factors combined with activity profiles for vessels calling at each port.

A.6.1. Data Sources

Consistent with the C3 RIA, the three primary C3 activity data sources used in this assessment were the US Army Corps of Engineers Entrances and Clearances data, Lloyd's Register of Ships (Lloyd's data), and Marine Exchange/Port Authority (MEPA) data. Each is described below.

A.6.1.1. US Army Corps of Engineers Entrances and Clearances

The USACE 2011 Entrances and Clearances data[37] were used to determine ship calls (trips or visits to a port). The Maritime Administration (MARAD) maintains the Foreign Traffic Vessel Entrances and Clearances database, which contains statistics on U.S. foreign maritime trade. USACE compiles these data to build a database which contains information on the vessel International Maritime Organization (IMO) number, type of vessel, commodities, weight, customs districts and ports, and origins and destinations of goods.

There are several limitations to using USACE Entrances and Clearances data. First, they do not contain any average time in mode or speed information, which is important when estimating emissions. Second, they only represent foreign cargo movements. Domestic cargo traffic (U.S. ships delivering cargo from one U.S. port to another U.S. port covered under the Jones Act[38]) are not always accounted for in the database. However, some trips made by Jones Act vessels are accounted for if they are carrying cargo from a foreign port to a U.S. port or from a U.S. port to a foreign port, since these are considered foreign cargo movements. At most ports, domestic commerce is carried out by Category 2 ships, although there are a few exceptions, particularly on the West Coast.[39] While Automatic Identification System (AIS) data can be used to determine domestic cargo trips, the processing of these data is time consuming and not within the scope of this assessment. Third, the Entrances and Clearances data do not always match MEPA data because the USACE Entrances and Clearances data include cargo movements from both public and private terminals at a port while the MEPAs usually only cover calls at public terminals. Port Authorities generally do not have jurisdiction over private terminals. Since the USACE Entrances and Clearances data account for over 90% of the emissions from Category 3 ships calling on U.S. ports,[40] these limitations should not have a substantial impact on the calculation of C3 emissions for this report.

A.6.1.2. Lloyd's Register of Ships

Lloyd's Register of Ships offers the largest database of commercially available maritime data in the world and is produced by IHS Global Limited.[41] The 2014 version used in this assessment has details on 180,000

---

[37] USACE, *Vessel Entrances and Clearances*. Available at: http://www.navigationdatacenter.us/data/dataclen.htm.

[38] Merchant Seaman Protection and Relief 46 USCS Appx § 688 (2002) Title 46. Appendix. Shipping Chapter 18.

[39] ICF International, *Inventory Contribution of U.S. Flagged Vessels*, June 2008.

[40] Merchant Seaman Protection and Relief 46 USCS Appx § 688 (2002) Title 46. Appendix. Shipping Chapter 18.

[41] Available at: http://www.sea-web.com.

vessels and 200,000 companies that own, operate, and manage them. Lloyd's data contain the following information on ship characteristics that are important for preparing detailed marine vessel inventories:

- Name (current and former)
- Ship Type
- Build Date
- Flag

- Deadweight tonnage (DWT)
- Vessel service speed[42]
- Main engine power, size, and configuration
- Limited data on Auxiliary engines

All data are referenced to both ship name (current or former) and IMO number. Only IMO number is a unique identifier for each ship. Lloyd's insures many of the OGVs on an international basis, and for these vessels, the data are quite complete. For other ships using a different insurance certification authority, some of the data are missing such as main engine power, size and configuration, and vessel service speed.

A.6.1.3. Marine Exchange/Port Authority Data

For this analysis, Marine Exchange/Port Authorities data were used to estimate hoteling and maneuvering times. As with the C3 RIA, much of the MEPA data came from a 1999 report[43] that described how to calculate marine vessel activity at deep sea ports and contained detailed port activities of eight deep sea ports. The detailed inventories were developed by obtaining ship call data from MEPA at the various ports for 1996 and matching the various ship calls to data from Lloyd's Maritime Information Services to provide ship characteristics. A 2002 port emission inventory for the Port of Houston[44] was also added to the list of "Typical Ports" in the 1999 report. The ports for which detailed inventories were developed are shown in Table A-13 along with the level of detail of shifts (movements within ports between berth and anchorage or between different berths) for each port. Most ports provided the ship name, IMO number, the vessel type, the date and time the vessel entered and left the port, and the vessel flag.

In addition to the detailed port inventories of ship activity, the 1999 report (which was also used in the C3 RIA) laid out a methodology to determine which of the typical ports can be used as a surrogate for other "like" or "modeled" ports that were being modeled in the C3 RIA. Consistent with the C3 RIA, hoteling and maneuvering times from the typical ports were used in this analysis to determine emissions at berth at the modeled ports. If the typical port data included shifts and anchorages, these data were also used in the modeled port. Anchorage times were only broken out for the Patapsco River Ports (Baltimore) in the typical port data.

---

[42] Vessel service speed is the average speed maintained by a ship under normal load and weather conditions.

[43] ARCADIS Geraghty & Miller, *Commercial Marine Activity for Deep Sea Ports in the United States,* EPA Report EPA420-R-99-020, September 1999. Available at http://www.epa.gov/otaq/models/nonrdmdl/c-marine/r99020.pdf .

[44] Starcrest Consulting Group, *Houston Galveston Area Vessel Emissions Inventory,* November 2002.

Table A-13. Deep Sea MEPA Vessel Movement and Shifting Details Used in the C3 RIA

MEPA Area and Ports[a]	Data Year	MEPA Data Include
Lower Mississippi River including the ports of New Orleans, South Louisiana, Plaquemines, and Baton Rouge	1996	Information on the first and last pier/wharf/dock (PWD) for the vessel (gives information for at most one shift per vessel). No information on intermediate PWDs, the time of arrival at the first destination PWD, or the time of departure from the River.
Consolidated Port of New York and New Jersey and other ports on the Hudson and Elizabeth Rivers	1996	All PWDs or anchorages for shifting are named. Shifting arrival and departure times are not given. Maneuvering and hoteling times are estimated from average speed and distance rather than calculated from date and time fields.
Delaware River Ports including the ports of Philadelphia, Camden, Wilmington and others	1996	All PWDs or anchorages for shifting are named. Shifting arrival and departure times are not given. Maneuvering and hoteling times are estimated from average speed and distance rather than calculated from date and time fields.
Puget Sound Area Ports including the ports of Seattle, Tacoma, Olympia, Bellingham, Anacortes, and Grays Harbor	1996	All PWDs or anchorages for shifting are named. Arrival and departure dates and times are noted for all movements, allowing calculation of maneuvering and hoteling both for individual shifts and the overall call on port.
The Port of Corpus Christi, TX	1996	Only has information on destination PWD and date and time in and out of the port area. No shifting details.
The Port of Coos Bay, OR	1996	Only has information on destination PWD and date and time in and out of the port area. No shifting details.
Patapsco River Ports including the port of Baltimore Harbor, MD	1996	All PWDs or anchorages for shifting are named. Shifting arrival and departure times are not given. Maneuvering and hoteling times are estimated from average speed and distance rather than calculated from date and time fields.
The Port of Tampa, FL	1996	All PWDs or anchorages for shifting are named. Arrival and departure dates and times are noted for all movements, allowing calculation of maneuvering and hoteling both for individual shifts and the overall call.
Port of Houston	2000	PWD and shifts provided. RSZ, maneuvering and hoteling times were calculated from available data by ship type.

[a] All marine exchange/port authority data listed above were for 1996 and were taken directly from the 1999 EPA Report. Port of Houston data were for 2000 and were taken directly for the Starcrest 2002 inventory report.[45]

Several emission inventories have been published since the original C3 RIA. They provided additional data on hoteling times, anchorage times, and maneuvering times and were used in this analysis in place of the data used in the C3 RIA, where appropriate. The updated data sources are listed in Table A-14. Some older inventories were also used to determine maneuvering and hoteling times, where newer data were not available. In general, maneuvering, hoteling and anchorage times were obtained from the

---

[45] Starcrest Consulting Group, *Houston Galveston Area Vessel Emissions Inventory*, November 2002.

1999 report unless detailed in Table A-14. The C3 RIA was used to determine RSZ speeds unless noted below.

Table A-14. More Recently Published Port Emission Inventories Since C3 RIA

Port Inventory	Data Year	Data Include
Port of New York & New Jersey Starcrest Consulting Group, *The Port of New York and New Jersey Port Commerce Department 2012 Multi-Facility Emissions Inventory*, August 2014. https://www.panynj.gov/about/pdf/panynj-multi-facility-ei-report_2012.pdf	2012	Information to calculate maneuvering times and hoteling times by ship type for all ship types
Puget Sound Area Ports Starcrest Consulting Group, *2011 Puget Sound Maritime Emission Inventory*, September 2012. http://www.pugetsoundmaritimeairforum.org/uploads/PV_FINAL_POT_2011_PSEI_Report_Update_23_May_13_scg.pdf	2011	Information to calculate maneuvering times and hoteling times by ship type for all ship types
Port of Charleston Moffatt & Nichol, *South Carolina State Ports Authority 2011 Emissions Inventory Update*, April 2013 http://www.scspa.com/SCSPA_Emission_Inventory_2011_Final_Report_1April2013.pdf	2011	Maneuvering and hoteling times by ship type for all ship types
Port of Virginia Moffatt & Nichol, *The Port of Virginia 2011 Comprehensive Emissions Inventory Update*, January 2013	2011	RSZ times by area and ship type for all ship types
Port of Baltimore Maryland Port Authority et al., *Emission Reductions from Port of Baltimore Maritime Vessels and Cargo Handling Equipment*, September 2010	2010	RSZ times by ship type for all ship types
Port of Houston Starcrest Consulting Group, *2007 Goods Movement Air Emissions Inventory at the Port of Houston*, January 2009 http://www.portofhouston.com/static/gen/inside-the-port/Environment/PHA-GM-AirEmissions-07.pdf	2007	RSZ speeds by area and ship type for all ship types
Port of Corpus Christi 2005 hoteling data obtained directly from Port of Corpus Christi Email from Sarah Kowalski, July 2007	2005	Hoteling time by ship type for all ships types. (These data replaced the typical port hoteling times for Port of Corpus Christi and were used for hoteling times for Port Arthur, Corpus Christi, and for tanker ships at Port of Houston.)

A.6.1.4. Port Matching

Both the original 1996 ports data shown in Table A-13 and more recently published data shown in Table A-14 were used as "typical ports". The inventory data from these typical ports were used as surrogates to estimate maneuvering time, hoteling time, and anchorage time (if available) for each of the 19 modeled ports unless noted in Table A-15 below. The typical port used in this analysis and those used in the C3 RIA are shown in the table. The typical ports that have changed since the C3 RIA are italicized and further explanation about why this change has been made is provided in Table A-16 below.

Table A-15. OGV Published Inventories and Data

Port Name	Published Inventory Data Used
Port of New York and New Jersey	*2012 Port of New York/New Jersey Inventory*
Port of New Orleans	1996 Lower Mississippi River Inventory
Port of Miami	1996 Port of Tampa for passenger ship hoteling time 1996 Delaware River Inventory for other ship types
Port of South Louisiana	1996 Lower Mississippi River Inventory
Port of Seattle	*2011 Puget Sound Inventory*
Port of Baton Rouge	1996 Lower Mississippi River Inventory
Port Arthur	*1996 Port of Corpus Christi Inventory for maneuvering* *2005 Port of Corpus Christi Inventory for hoteling*
Port of Portland	*2012 New York/New Jersey Inventory*
Port of Mobile	*1996 Delaware River Inventory*
Port of Houston	*2007 Houston Inventory for RSZ speeds* *2000 Houston Inventory used for maneuvering times* *2005 Corpus Christi data for Tanker hoteling time* *2012 New York/New Jersey Inventory for other ship hoteling times*
Port of Baltimore	*2010 Port of Baltimore Report for RSZ speeds* *1996 Patapsco River Inventory for maneuvering, hoteling and anchorage*
Port of Hampton Roads (Norfolk)	*2011 Port of Virginia Inventory for RSZ speeds* *1996 Patapsco River Inventory for maneuvering, hoteling and anchorage*
Port of Philadelphia	1996 Delaware River Inventory
Port of Charleston	*2011 Port of Charleston Inventory*
Port of Corpus Christi	*1996 Port of Corpus Christi Inventory for maneuvering* *2005 Port of Corpus Christi Inventory for hoteling*
Port Tampa Bay	*1996 Port of Tampa for passenger ship hoteling time* *1996 Delaware River Inventory for other ship types*
Port of Savannah	1996 Patapsco River Inventory
Port of Coos Bay, OR	1996 Port of Coos Bay Inventory
Port of San Juan, PR	*1996 Delaware River Typical Port Data*

Notes: Italics denote ports that use a different matching typical port than what was used in the C3 RIA.
Year given reflects data year, not report year.

Table A-16. Typical Port Matching

Port Name	Current Matching Typical Port	C3 RIA Matching Typical Port
Port of New York and New Jersey	*2012 Port of New York/New Jersey Inventory*	*1996 Port of New York/New Jersey Inventory*
Port of New Orleans	1996 Lower Mississippi River Inventory	1996 Lower Mississippi River Inventory
Port of Miami	1996 Port of Tampa for passenger ship hoteling time 1996 Delaware River Inventory for other ship types	1996 Delaware River Inventory
Port of South Louisiana	1996 Lower Mississippi River Inventory	1996 Lower Mississippi River Inventory
Port of Seattle	*2011 Puget Sound Inventory*	*1996 Puget Sound Inventory*
Port of Baton Rouge	1996 Lower Mississippi River Inventory	1996 Lower Mississippi River Inventory
Port Arthur	*1996 Port of Corpus Christi Inventory for maneuvering* *2005 Port of Corpus Christi Inventory for hoteling*	*1996 Port of Corpus Christi Inventory*
Port of Portland	*2012 New York/New Jersey Inventory*	*1996 Puget Sound Inventory*
Port of Mobile	*1996 Delaware River Inventory*	*1996 Port of Corpus Christi Inventory*
Port of Houston	*2007 Houston Inventory for RSZ speeds* *2000 Houston Inventory used for maneuvering times* *2005 Corpus Christi data for Tanker hoteling time* *2012 New York/New Jersey Inventory for other ship hoteling times*	*2000 Port of Houston Inventory*
Port of Baltimore	*2010 Port of Baltimore Report for RSZ speeds* *1996 Patapsco River Inventory for maneuvering, hoteling and anchorage*	*1996 Patapsco River Inventory*
Port of Hampton Roads (Norfolk)	*2011 Port of Virginia Inventory for RSZ speeds* *1996 Patapsco River Inventory for maneuvering, hoteling and anchorage*	*Not in C3 RIA*
Port of Philadelphia	1996 Delaware River Inventory	1996 Delaware River Inventory
Port of Charleston	*2011 Port of Charleston Inventory*	*1996 Delaware River Inventory*
Port of Corpus Christi	*1996 Port of Corpus Christi Inventory for maneuvering* *2005 Port of Corpus Christi Inventory for hoteling*	*1996 Port of Corpus Christi Inventory*
Port Tampa Bay	*1996 Port of Tampa for passenger ship hoteling time* *1996 Delaware River Inventory for other ship types*	*1996 Port of Tampa Inventory*
Port of Savannah	1996 Patapsco River Inventory	1996 Patapsco River Inventory
Port of Coos Bay, OR	1996 Port of Coos Bay Inventory	1996 Port of Coos Bay Inventory
Port of San Juan, PR	*1996 Delaware River Typical Port Data*	*Not in C3 RIA*

Notes: Italics denote ports that use a different matching typical port than what was used in the C3 RIA.
Year given reflects data year, not report year.

For many of the ports listed in Table A-16, the data were from 1996. However, these data were used to determine operating characteristics, and it is expected that ship operations at the ports should not have significantly changed since 1996 unless there was a significant change in the physical layout of the port.

While the Port of Hampton Roads (Norfolk) was not part of the C3 RIA, Port of Newport News, which is nearby, used the 1996 Patapsco River Inventory as the matching typical port. Because the Port of Newport News was included in the C3 RIA and is in the same waterway, the 1996 Patapsco River Inventory was used as the matching typical port for Hampton Roads.

Other ports were matched to typical ports based on the type of ships that frequent the port and port cargo throughput. For example, a majority of the ship calls at the Port of Portland and San Juan are container ships while at Port Arthur the majority is tankers. Thus, Portland and San Juan were matched with other container ports and Port Arthur with other tanker ports. The Port of Mobile is served by a variety of vessel types more consistent with the Delaware River ports, than with Corpus Christi, which only covered tanker vessels. The Port of Houston has both container and tanker traffic, so two different typical ports were used for hoteling times. Newer 2005 Corpus Christi data were used for tanker hoteling for the Port of Houston as well as for all ships at Port Arthur. These newer port data provide updated information over the 1996 data. The 1996 Delaware River Inventory was also used for Port Tampa Bay because the original 1996 Port of Tampa Inventory no longer reflected the type of vessels operating at Port Tampa Bay. For example, in 1996, there were only four containerships visiting the port and in 2011, there were 79 containerships. As a result, the Delaware Inventory was believed to be a better match. However, the 1996 Port of Tampa inventory provided much better hoteling time information for passenger ships for both Tampa and Miami than the Delaware inventory since both those ports have high passenger ship volumes as opposed to the Delaware River ports.

The remaining container ports that use different matching ports in this analysis from the C3 RIA (as shown in Table A-16) were matched based upon TEUs unloaded and loaded per call. Amount of TEUs being unloaded are one factor that affects hoteling time, another important factor is how efficient unloading operations are at a port. Evaluating unloading operations and other factors that may affect hoteling time were beyond the scope of this analysis, and TEUs were used as a proxy for hoteling time. TEUs by port and calls were used to match maneuvering and hotelling times to typical ports. The 2012 Port of New York/New Jersey was used as a matching typical port for maneuvering and hoteling times for the Port of Portland and for non-tanker ships for Houston. TEUs per call for the Ports of Mobile, Miami, Tampa Bay,[46] and San Juan were similar to the Port of Philadelphia (a Delaware River Typical Port), so the 1996 Delaware River Inventory data were used as the typical port for those ports.

Generally, the typical port data show different hoteling and maneuvering times based upon ship size. In some cases, the specific vessel deadweight tonnage (DWT) range bin at the modeled port was not in the typical port data. In those cases, the next nearest DWT range bin was used for the calculations, which was done in the C3 RIA. In a few cases, the engine type for a given ship type at the modeled port might not be in the

---

[46] While Tampa was one of the original typical ports used in the C3 RIA, it provides minimal container ship data and thus the 1996 Delaware River typical port data were substituted.

typical port data. In these cases, the closest engine type at the typical port was used. Also in a few cases, a specific ship type in the modeled port data was not in the typical port data. In this case, data from the most similar ship type at the typical port were used at the modeled port. Section A.6.7 provides more information on bin matching.

A.6.2. Modeling Approach

Using the C3 RIA modeling approach, the OGV emission inventories were calculated using energy-based emission factors combined with activity profiles for each vessel. Ships calling on each port were binned by ship type, engine type, and DWT, and emissions for each mode were calculated for auxiliary and main propulsion engines using the following general equation:

$$E = P \times LF \times A \times EF \qquad \text{Eq. A-1}$$

Where

E = Emissions (grams [g]),

P = Maximum Continuous Rating Power (kilowatts [kW]),

LF = Load Factor (percent of vessel's total power),

A = Activity (hours [h]) (hours/call * # of calls), and

EF = Emission Factor (grams per kilowatt-hour [g/kWh]).

The emission factor is in terms of emissions per unit of energy from the engine. It is multiplied by the energy needed to move the ship or perform other activities like hoteling. Only four modes of activity were considered in this assessment: reduced speed zone, maneuvering, hoteling, and anchorage (if available). Cruising emissions were omitted since the scope of this assessment is limited to near-port emissions.

OGVs vary greatly in design, operating speeds, and engine sizes based on vessel type. For this analysis, vessel types were broken out by the cargo they carry. Table A-17 lists the OGV types that are used in this analysis.

Table A-17. Oceangoing Vessel Ship Types

Ship Type	Description
Auto Carrier	Self-propelled dry-cargo vessels that carry containerized automobiles
Bulk Carrier*	Self-propelled dry-cargo ship that carries loose cargo
Container Ship	Self-propelled dry-cargo vessel that carries containerized cargo
General Cargo	Self-propelled cargo vessel that carries a variety of dry cargo
Miscellaneous	Category for those vessels that do not fit into one of the other categories or are unidentified
Passenger	Self-propelled cruise ships
Reefer	Self-propelled dry-cargo vessels that often carry perishable items
Roll-on/Roll-off (RORO)	Self-propelled vessel that handles cargo that is rolled on and off the ship, including ferries
Tanker	Self-propelled liquid-cargo vessels including chemical tankers, petroleum product tankers, liquid food product tankers, etc.
Tug	Self-propelled tugboats and towboats that tow/push cargo or barges in the open ocean

*In the 2002 C3 analysis, barge carriers were considered as a separate vessel type. Due to the small number of barge carriers in the dataset, barge carriers are considered as bulk carriers.

Other characteristics that are determined from Lloyd's data are the build year, propulsion engine power, vessel service speed, engine type (i.e., diesel, gas or steam turbines), and vessel DWT. Vessel service speed is the average speed maintained by a ship under normal load and weather conditions in the open ocean. Design speed is the maximum speed a vessel can travel but is generally not used due to high fuel consumption requirements.

Both propulsion and auxiliary marine vessel engines are defined by the categories shown in Table A-18. Note that only vessels with Category 3 propulsion engines are considered in the OGV sector. Emissions from Category 1 and 2 vessels are addressed in the harbor craft sector. Most ships have diesel engines, although some ships have steam turbines (ST) and others have gas turbines (GT). Some ships are electric drive (ED). For the purpose of this analysis, Category 3 slow speed diesel (SSD) are considered 2-stroke engines while Category 3 medium speed diesel (MSD) engines are considered 4-stroke engines. Engine speed designations for diesel powered ships are shown in Table A-19.

Table A-18. EPA Marine Compression Ignition Engine Categories

Category	Specification	Use
1	Gross Engine Power ≥ 37 kW* Displacement < 7 liters per cylinder	Small harbor craft and recreational propulsion
2	Displacement ≥ 7 and < 30 liters per cylinder	OGV auxiliary engines, harbor craft, and smaller OGV propulsion
3	Displacement ≥ 30 liters per cylinder	OGV propulsion

* EPA assumes that all engines with a gross power below 37 kW are used for recreational applications and are treated separately from the commercial marine category.

Table A-19. Marine Diesel Engine Speed Designations

Speed Category	Engine Stroke Type	Category
SSD	2	3
MSD	4	2, 3
HSD	4	1

Information on auxiliary engine power is generally not complete in the Lloyd's data. As was done in the C3 RIA, auxiliary engine power was estimated using a survey performed by California Air Resources Board (ARB) in 2005.[47] The survey provided average propulsion and auxiliary engine power by ship type and is shown in Table A-20. Auxiliary to propulsion power ratios were calculated for each ship type from the survey and used to calculate auxiliary power for all ships. All auxiliary power is assumed to be provided by Category 2 medium speed diesel engines, consistent with the C3 RIA methodology.

---

[47] California Air Resources Board, *2005 Oceangoing Ship Survey, Summary of Results,* September 2005

Table A-20. Auxiliary Engine Power Ratios (ARB Survey, Except as Noted)

Ship Type	Average Propulsion Engine (kW)	Average Auxiliary Engines				Auxiliary to Propulsion Ratio
		Number	Power Each (kW)	Total Power (kW)	Engine Speed	
Auto Carrier	10,700	2.9	983	2,850	MSD	0.266
Bulk Carrier	8,000	2.9	612	1,776	MSD	0.222
Container Ship	30,900	3.6	1,889	6,800	MSD	0.220
Passenger Ship[a]	39,600	4.7	2,340	11,000	MSD	0.278
General Cargo	9,300	2.9	612	1,776	MSD	0.191
Miscellaneous[b]	6,250	2.9	580	1,680	MSD	0.269
RORO	11,000	2.9	983	2,850	MSD	0.259
Reefer	9,600	4.0	975	3,900	MSD	0.406
Tanker	9,400	2.7	735	1,985	MSD	0.211

[a] Passenger ships typically use a different engine configuration known as diesel-electric. These vessels use large generator sets for both propulsion and ship-board electricity. The figures for passenger ships above are estimates taken from the Starcrest Vessel Boarding Program and were used for modeling purposes to split emissions between propulsion and auxiliary use.

[b] Miscellaneous ship types were not provided in the ARB methodology, so values from the Starcrest Vessel Boarding Program were used. All tugs are considered as miscellaneous.

A.6.3. Emission Boundaries

The emission boundaries for the OGV sector were defined as any activity occurring within 5 km of the National Emissions Inventory port shape files,[48] with the following exceptions:

- New Orleans, Baton Rouge, and South Louisiana: The boundaries for these ports were extended by 0.5 km beyond the port shape files because of the expected impact RSZ emissions would have on the individual ports and to account for the RSZ emissions at the port boundaries.

- Port of Norfolk: The boundary for this port includes all activity (except that going to Newport News) within 10 km of the Norfolk port shape files. The boundary was extended to 10 km to capture the RSZ emissions near residential areas that were identified as important for this analysis. One RSZ length was used for all ships.

- Port of Savannah: The boundary for this port is 10 km towards the ocean from the port shape. The boundary was extended to 10 km to capture the RSZ emissions near residential areas. It is truncated at the north end of the port.

- Port of Philadelphia: While the 5 km boundary includes the Port of Camden, hoteling and maneuvering emissions from Camden were beyond the scope of this assessment and are not considered here.

---

[48] U.S. EPA, *2011 National Emissions Inventory Port Shape Files,* August 2014. Available at: http://www.epa.gov/ttn/chief/eis/2011nei/ports_20140729.zip

A.6.4. Activity Modes

The following activity measurements and modes were analyzed in this assessment: number of calls, reduced speed zones, maneuvering, hoteling, and anchorage.

A.6.4.1. Calls

The number of calls by C3 vessels at each port was determined from the USACE Entrances and Clearances data. As was done in the C3 RIA, barges were removed from the data as these are non-propelled vessels generally moved by a Category 1 or 2 tug. It is important to note that Entrances and Clearances only record stops where foreign cargo was either loaded or discharged (i.e., where a ship discharges goods but doesn't load goods, only the entrance was recorded). Thus, to get a better estimate of calls, the maximum of either entrances or clearances for a given ship type, engine type, and DWT bin is used in this analysis as a surrogate for calls. This approach differs slightly from the C3 RIA methodology, which defined calls as the average of entrances and clearances for a given ship type, engine type, and DWT bin.

A.6.4.2. Reduced Speed Zone

The reduced speed zone activity in this analysis included movements from the port entrance to the boundary of the analysis. Port entrances were determined from the National Emission Inventory (NEI) shapefiles for the port. RSZ distances were measured using Google Earth. An example of this is shown in Figure A-2, where the port shape in orange defines the port. The red line depicts the port boundary and the yellow line defines the RSZ shipping lane within the boundary.

Figure A-2. Savannah RSZ Measurement

Table A-21 lists the RSZ distances and speeds for each port. For the Ports of Houston and Philadelphia, "fast" ships are defined as auto carriers, container ships, passenger ships and ROROs. In addition, the RSZ speed for the Port of Savannah is 13 knots; however, the cruise speed for some ships entering that port is less than 13. The analysis for these ships used their cruise speed instead of the RSZ speed. The Port of Houston RSZ is a special case as has two reduced speed zones to accommodate different speeds within the Houston Ship Channel. Details for these calculations may be found in section A.6.9.

Table A-21. RSZ One-Way Distances and Speeds

Port Name	Distance (km)	Speed (knots)	RSZ Speed Source
Port of New York and New Jersey	5.0	Containers – 10.0 Tankers – 6.2 Others – 8.0	2012 Port of New York/New Jersey Inventory
Port of New Orleans	28.2	10.0	C3 RIA
Port of Miami	5.0	12.0	C3 RIA
Port of South Louisiana	44.0	10.0	C3 RIA
Port of Seattle	5.6	Container – 14.0 Passenger – 15.0 Others – 8.0	2011 Puget Sound Inventory
Port of Baton Rouge	71.3	10.0	C3 RIA
Port Arthur	5.6	7.0	C3 RIA
Port of Portland	4.6	8.4	C3 RIA
Port of Mobile	5.0	11.0	C3 RIA
Port of Houston	14.1 / 10.4	Fast Ships – 12.0 Slow Ships – 9.5	2007 Port of Houston Inventory
	16.7	Fast Ships – 10.5 Slow Ships – 7.5	
Port of Baltimore	7.9	Auto Carrier – 16.5 Bulk Carrier – 12.9 Container Ship – 17.5 General Cargo – 13.3 Passenger Ship – 17.0 RoRo – 14.9 Tanker – 13.0	2010 Port of Baltimore Report
Port of Norfolk (Hampton Roads)	18.1	Container – 11.4 Other ships – 10.0	2011 Port of Virginia Inventory
Port of Philadelphia	5.5	Fast Ships – 11.0 Slow Ships – 9.0	C3 RIA
Port of Charleston	7.4	12.0	C3 RIA
Port of Corpus Christi	5.0	< 90,000 DWT – 12.0 > 90,000 DWT – 9.0	C3 RIA
Port Tampa Bay	5.0	9.0	C3 RIA
Port of Savannah	12.0	13.0	C3 RIA
Port of Coos Bay, OR	7.6	6.5	C3 RIA
Port of San Juan, PR	5.0	10.0	PR&VI ECA[49]

[49] U.S. Environmental Protection Agency, *Proposal to Designate an Emission Control Area for Nitrogen Oxides, Sulfur Oxides and Particulate Matter*, Technical Support Document, Report EPA-420-R-10-013, August 2010. Available at http://www.epa.gov/otaq/regs/nonroad/marine/ci/420r10013.pdf.

A.6.4.3. Maneuvering

Ships typically transition from RSZ to maneuvering speed as they begin to approach their berth. The maneuvering distance is different for each port but is generally defined as slow speed activity within a port. This can include movement from the port entrance to the berth, docking at the berth, and any intra-port shifts that are shown in the matching typical port inventory. For purposes of computing propulsion load factors and applying low speed adjustment factors, maneuvering is considered to occur at stall speed (5.8 knots), which is consistent with the C3 RIA. Actual speeds are less due to starting, stopping, and reversing that occur during maneuvering. In addition, most ships use assist tugs to push them into and out of a berth.

Consistent with the C3 RIA, maneuvering times were taken from the matching typical port data. For Port of New York/New Jersey, Port of Seattle, and Port of Charleston, maneuvering information from their newer port inventories listed in Table A-15 were used. Details for these calculations may be found in section A.6.9.

A.6.4.4. Hoteling

Hoteling is the time at berth when the vessel is operating auxiliary engines only or is using shore power. Except when the vessel is using shore power, auxiliary engines are operating under load the entire time the vessel is manned. Peak loads occur after the propulsion engines are shut down, when the auxiliary engines are responsible for all onboard power and/or are used to power off-loading equipment. Hoteling activity needs to be divided into auxiliary engine use and shore power to accurately account for reduced emissions from shore power. No shore power was considered in the C3 RIA.

As with the C3 RIA, average hoteling times by ship type and DWT range were taken directly from the matching typical port data. For Port of New York/New Jersey, Port of Seattle, and Port of Charleston, hoteling information from their newer port inventories listed in Table A-15 were used. Additionally, Port of Corpus Christi was updated with newer hoteling information provided by the port. Details for these calculations may be found in section A.6.9.

A.6.4.5. Anchorage

Anchorage occurs when the ship has arrived at a port but no berth is available. Anchorage is only considered if it occurs within the port boundaries and anchorage data are available. Shore power and advanced marine emission control systems cannot be applied while at anchorage. While the C3 RIA assumed all calls involve anchorage, it was assumed in this assessment that not all ship calls involve anchorage. To estimate the percentage of calls in which anchorage occurs in the 1996 Patapsco River Inventory, the data were reanalyzed to provide that percentage by ship type/engine type/DWT bin. This represents the number of calls for which a ship type/engine type/DWT bin anchors within the port (but not at a berth) divided by the total calls for that bin. Anchorage time was calculated accordingly. This is consistent with the newer Starcrest methodology used for Port of Los Angeles[50] and Long Beach.[51]

A.6.4.6. Summary

The various activity modes for Category 3 vessels analyzed in this assessment are summarized in Table A-22.

Table A-22. Vessel Movements and Time-In-Mode Descriptions

Summary Table Field	Description
Call	A call is one entrance and one clearance. Since the USACE Entrances and Clearances data do not provide a record for an entrance where no foreign cargo discharged or a record of a clearance where no foreign cargo is loaded at a port, the number of entrances and clearances may not be the same. Therefore, the number of calls were taken as the maximum of the entrances or clearances at a port as grouped by ship type, engine type, and deadweight tonnage bin.
Reduced Speed Zone (RSZ) (hr/call)	Time when a ship reduces speed before entering a port. This can be a long distance down a river or channel and generally ends at the port entrance.
Maneuver (hr/call)	Time when a ship is being berthed or de-berthed, traveling to an anchorage or moving between berths. Maneuvering is assumed to occur within the port area, generally beginning and ending at the entrance of the port. This will include shifts within a port area moving from one berth to another. For purposes of calculating load factors, maneuvering was assumed to occur at an average speed of 5.8 knots. Maneuvering times were taken from the typical port data or calculated from published inventories.
Hoteling (hr/call)	Hoteling is the time at berth when the vessel is operating auxiliary engines only or is using shore power. Peak loads occur after the propulsion engines are shut down, when the auxiliary engines or shore power is responsible for all onboard power and/or are used to power off-loading equipment.
Anchorage (hr/call)	If the port data included anchorage, it is broken out separately for this analysis. Some emission reduction techniques cannot be applied while at anchorage. This mode was ignored if not specifically identified.

A.6.5. Load Factors

As in the C3 RIA, load factors are expressed as a percent of the vessel's total propulsion or auxiliary power. At service or cruise speed, the propulsion load factor is assumed to be 83%. At lower speeds, the Propeller Law is used to estimate ship propulsion loads, based on the theory that propulsion power varies by the cube of speed as shown in the equation below.

---

[50] Starcrest Consulting Group, *Port of Los Angeles Air Emissions Inventory – 2011*, July 2012.
[51] Starcrest Consulting Group, *Port of Long Beach Air Emissions Inventory – 2011*, July 2012.

$$LF = (AS/MS)^3 \hspace{4cm} \text{Eq. A-2}$$

Where

LF = Load Factor (percent),

AS = Actual Speed (knots), and

MS = Maximum Speed (knots).

Maximum speed is calculated from service speed, which is available in the Lloyd's data, as 1.00/0.94 multiplied by service speed as was done in the C3 RIA. While load factors were calculated using the above propeller law, load factors below 2% were set to 2% as a minimum.[52]

Load factors for auxiliary engines vary by ship type and operating mode. Table A-23 shows the auxiliary engine load factors determined by Starcrest, through interviews conducted with ship captains, chief engineers, and pilots during its vessel boarding programs.[53] These were used in the C3 RIA. Auxiliary load factors listed in Table A-23 are used together with the total auxiliary engine power (determined from total propulsion power and the ratios from Table A-20) to calculate auxiliary engine emissions. Consistent with the C3 RIA, emissions from auxiliary boilers are not explicitly calculated here, but the load factors presented below are large enough to include emissions from both auxiliary engines and auxiliary boilers.

Table A-23. Auxiliary Engine Load Factor Assumptions

Ship Type	RSZ	Maneuver	Hotel Anchor
Auto Carrier	0.30	0.67	0.24
Bulk Carrier	0.27	0.45	0.22
Container Ship	0.25	0.50	0.17
Passenger Ship	0.80	0.80	0.64
General Cargo	0.27	0.45	0.22
Miscellaneous	0.27	0.45	0.22
RORO	0.30	0.45	0.30
Reefer	0.34	0.67	0.34
Tanker	0.27	0.45	0.67
Tug	0.27	0.45	0.22

---

[52] Starcrest Consulting Group LLC, *Update to the Commercial Marine Inventory for Texas to Review Emission Factors, Consider a Ton-Mile EI Method, and Revised Emissions for the Beaumont-Port Arthur Non-Attainment Area*, prepared for the Houston Advanced Research Center, January 2004.

[53] Starcrest Consulting Group, *Port-Wide Baseline Air Emissions Inventory*, prepared for the Port of Los Angeles, June 2004.

A.6.6. Emission Factors

Emission factors vary by engine type, engine tier, fuel type, fuel sulfur levels, and load factor, for both propulsion and auxiliary engines.

A.6.6.1. Propulsion Engine Emission Factors

Propulsion engine emission factors used in this analysis are shown in Table A-24. They come from a European Commission study (referred to here as Entec)[54] and are similar to the ones used in the C3 RIA. For example, the brake specific fuel consumption (BSFC) for MSD in this analysis was 213 g/kWh for residual oil (RO) and 203 g/kWh for marine distillate oil/marine gas oil (MDO/MGO) compared to the 210 g/kWh and 200 g/kWh, respectively, used in the C3 RIA. PM_{10} emission factors for MSD engines were also recalculated using the equations listed below that were determined based on existing engine data in consultation with ARB.

$$\text{RO } PM_{10} \text{ EF} = 1.35 + \text{BSFC} \times 7 \times 0.02247 \times (\text{Fuel Sulfur Fraction} - 0.0246) \qquad \text{Eq. A-3}$$

$$\text{MDO \& MGO } PM_{10} \text{ EF} = 0.23 + \text{BSFC} \times 7 \times 0.02247 \times (\text{Fuel Sulfur Fraction} - 0.0024) \qquad \text{Eq. A-4}$$

$$PM_{2.5} \text{ EF} = 0.92 \times PM_{10} \text{ EF} \qquad \text{Eq. A-5}$$

The above equations are based upon the fact that the sulfate component in PM_{10} has a molecular weight seven times that of sulfur and the assumption that 2.247% of the fuel sulfur is converted to PM_{10} sulfate. $PM_{2.5}$ was assumed to be 92% of PM_{10}. These assumptions and formulas were used in the C3 RIA.

For SO_2, the emission factors were based upon a fuel sulfur to SO_2 conversion factor from ENVIRON, assuming that 97.753% of the fuel sulfur was converted to SO_2 and taking into account the molecular weight difference between SO_2 and sulfur.[55]

$$SO_2 \text{ EF} = \text{BSFC} \times 2 \times 0.97753 \times \text{Fuel Sulfur Fraction} \qquad \text{Eq. A-6}$$

CO_2 emission factors were calculated from the BSFC assuming a fuel carbon content of 86.7% by weight[56] and a ratio of molecular weights of CO_2 and C at 3.667.

$$CO_2 \text{ EF} = \text{BSFC} \times 0.867 \times 3.667 \qquad \text{Eq. A-7}$$

---

[54] Entec UK Limited, *Quantification of Emissions from Ships Associated with Ship Movements between Ports in the European Community*, prepared for the European Commission, July 2002.

[55] Memo from Chris Lindhjem of ENVIRON, *PM Emission Factors,* December 15, 2005.

[56] Entec UK Limited, *Quantification of Emissions from Ships Associated with Ship Movements between Ports in the European Community*, prepared for the European Commission, July 2002.

Table A-24. Emission Factors for Tier 0 OGV Propulsion Engines, g/kWh

Engine Type	Fuel Type	Fuel Sulfur	Emission Factors (g/kWh)							
			NOx	PM$_{10}$	PM$_{2.5}$	HC	CO	SO$_2$	CO$_2$	BSFC
SSD	RO	2.70%	18.1	1.42	1.31	0.6	1.4	10.29	621	195
	MGO/MDO	0.10%	17.0	0.19	0.17	0.6	1.4	0.36	589	185
MSD	RO	2.70%	14.0	1.43	1.32	0.5	1.1	11.24	678	213
	MGO/MDO	0.10%	13.2	0.19	0.17	0.5	1.1	0.40	646	203
GT	RO	2.70%	6.1	1.47	1.35	0.1	0.2	16.10	971	305
	MGO/MDO	0.10%	5.7	0.17	0.15	0.1	0.2	0.57	923	290
ST	RO	2.70%	2.1	1.47	1.35	0.1	0.2	16.10	971	305
	MGO/MDO	0.10%	2.0	0.17	0.15	0.1	0.2	0.57	923	290

The IMO adopted mandatory Tier I NOx emission limits in Annex VI to the International Convention for Prevention of Pollution from Ships in 1997. These NOx limits apply for all marine engines over 130 kW for engines built on or after January 1, 2000, including those engines that underwent a major rebuild after January 1, 2000. For the C3 RIA, the effect of the IMO standard was determined to be a reduction in NOx emission rate of 11% below that for engines built before 2000. Therefore, for engines built between 2000 and 2010 (Tier I), a factor of 0.89 was applied to the calculation of NOx emissions for both propulsion and auxiliary engines.[57]

IMO Tier II NOx emission standards started in 2011. For the C3 RIA, the effect of Tier II was determined to be a NOx reduction of 2.5 g/kWh reduction over Tier I engines. Tier III took effect in 2016 and EPA determined in the C3 RIA that the effect of Tier III to be an 80% reduction from Tier I. Thus Tier III emission factors are 20% of Tier I emission factors. All emission factors used here are consistent with the C3 RIA.

In addition to the MARPOL Annex VI emission limits that apply to all ships engaged in international transportation, U.S. vessels must also comply with EPA's Clean Air Act requirements for engines and fuels. The NOx emission limits for Category 3 engines are equivalent to the MARPOL Annex VI NOx limits. EPA's sulfur limit for distillate locomotive or marine (LM) diesel fuel sold in the United States is more stringent than the ECA fuel sulfur limit; the sulfur limit for ECA fuel for use on Category 3 marine vessels is equivalent to the MARPOL Annex VI SOx limits. EPA also has standards for C3 engines,[58] which are generally the same or more stringent but almost all C3 engines used in international shipping fall under IMO regulations.

---

[57] In addition, as part of the IMO standards, marine diesel engines built between 1990 and 1999 that are ≥90 liters per cylinder need to be retrofit to meet Tier I emission standards upon engine rebuild if a retrofit kit is available. Consistent with the C3 RIA, it was assumed that 80% of these ships will have retrofit kits available and that this phase-in will happen over five years, with 20% of eligible ships each year starting in 2011. Since the 2011 phase-in represents less than 0.4% of NOx emissions by ships at the 19 ports, no engines were assumed to be rebuilt in the 2011 inventory for calculation purposes.

[58] U.S. Environmental Protection Agency, *Control of Emissions from New Marine Compression-Ignition Engines at or Above 30 Liters per Cylinder,* Federal Register, Vol 75, No 83, April 30, 2010.

As with the C3 RIA, emission factors are considered to be constant down to about 20% load. Below that threshold, emission factors tend to increase as the load decreases. This is because diesel engines are less efficient at low loads and the BSFC tends to increase. Thus, while mass emissions (grams per hour) decrease with low loads, the engine power tends to decrease more quickly, thereby increasing the emission factor (grams per engine power) as load decreases. Energy and Environmental Analysis Inc. demonstrated this effect in a study prepared for EPA in 2000.[59]

Low-load multiplicative adjustment factors used in the C3 RIA are presented in Table A-25. As these adjustment factors were derived for diesel engines, the low load adjustment factors should only be applied to MSD and SSD engines. This is a modification of the C3 RIA methodology where low load adjustment factors were also applied to steam turbine engines. However, since the boiler that drives the steam turbines also drives the auxiliary engines, the total load on the boiler (propulsion and auxiliary load) is higher than that the propulsion steam turbine load only. Thus, it is assumed that they are always higher than 20% and therefore no low load factor should be applied.

Table A-25. Calculated Propulsion Engine Low Load Multiplicative Adjustment Factors

Load	NOx	HC	CO	PM	SO$_2$	CO$_2$
2%	4.63	21.18	9.68	7.29	3.36	3.28
3%	2.92	11.68	6.46	4.33	2.49	2.44
4%	2.21	7.71	4.86	3.09	2.05	2.01
5%	1.83	5.61	3.89	2.44	1.79	1.76
6%	1.60	4.35	3.25	2.04	1.61	1.59
7%	1.45	3.52	2.79	1.79	1.49	1.47
8%	1.35	2.95	2.45	1.61	1.39	1.38
9%	1.27	2.52	2.18	1.48	1.32	1.31
10%	1.22	2.20	1.96	1.38	1.26	1.25
11%	1.17	1.96	1.79	1.30	1.21	1.21
12%	1.14	1.76	1.64	1.24	1.18	1.17
13%	1.11	1.60	1.52	1.19	1.14	1.14
14%	1.08	1.47	1.41	1.15	1.11	1.11
15%	1.06	1.36	1.32	1.11	1.09	1.08
16%	1.05	1.26	1.24	1.08	1.07	1.06
17%	1.03	1.18	1.17	1.06	1.05	1.04
18%	1.02	1.11	1.11	1.04	1.03	1.03
19%	1.01	1.05	1.05	1.02	1.01	1.01
20%	1.00	1.00	1.00	1.00	1.00	1.00

Low load adjustment factors were not applied to diesel electric drive systems for loads below 20% because in these systems, multiple engines are used to generate power, and some can be shut down to allow others to operate at a more efficient setting.

---

[59] Energy and Environmental Analysis Inc., *Analysis of Commercial Marine Vessels Emissions and Fuel Consumption Data*, EPA420-R-00-002, February 2000.

A.6.6.2. Auxiliary Engine Emission Factors

As with propulsion engines, the auxiliary engine emission factors used in this analysis comes from Entec[60] and was used in the C3 RIA. However, the BSFC used in this analysis differs from what was used in the C3 RIA. The BSFC was updated in this analysis to match the 227 g/kWh and 217 g/kWh for RO and MGO/MDO, respectively, to match that listed by Entec for auxiliary engines. This is instead of the 210 g/kWh and 200 g/kWh, respectively, that was used in the C3 RIA.

An ARB survey published in 2005[61] found that almost all ships used RO in their main propulsion engines. Only 29% of all ships (except passenger ships) used distillate (MGO/MDO) in their auxiliary engines, with the remaining 71% using RO. Only 8% of passenger ships used distillate in their auxiliary engines, while the other 92% used RO.

Distillate fuels discussed in the ARB survey ranged from 0.03% to 1.5% sulfur in those ships that used distillate in their auxiliaries. For the purposes of this analysis, the use of 1.0% sulfur distillate fuel in these engines was assumed for the 2011 baseline estimate. The C3 RIA assumed 1.5% sulfur distillate but calculated that the difference between using 1.0% versus 1.5% did not result in significant overestimation of emissions. The value used in this analysis is consistent with that used by Entec for global distillate sulfur levels. Note that no distinction was made here between MGO and MDO, and they are referred to here as "MGO/MDO."

The equations listed above in section A.6.6.1 were used to calculate the PM_{10}, $PM_{2.5}$, SO_2, and CO_2 emission factors for this analysis based upon the different fuel sulfur levels and BSFCs as mentioned above. All other factors match the C3 RIA values. Table A-26 provides these auxiliary engine emission factors. Consistent with the C3 RIA assumptions, there is no need for a low load adjustment factor for auxiliary engines, because they are generally operated in banks. When only low loads are needed, one or more engines are shut off, allowing the remaining engines to operate at a more efficient level.

Table A-26. Tier 0 Auxiliary Engine Emission Factors by Fuel Type, g/kWh

Fuel Type	Fuel Sulfur	Emission Factors (g/kWh)							
		NOx	PM_{10}	$PM_{2.5}$	HC	CO	SO_2	CO_2	BSFC
RO	2.70%	14.7	1.44	1.32	0.4	1.10	11.98	723	227
MGO/MDO	1.00%	13.9	0.49	0.45	0.4	1.10	4.24	668	217
MGO/MDO	0.10%	13.9	0.18	0.17	0.4	1.10	0.42	691	217

Using the percentages of RO and Distillate found in the ARB survey for passenger ships and other ships, Table A-27 provides weighted emission factors for the two ship types for use in the analyses.

---

[60] Entec UK Limited, *Quantification of Emissions from Ships Associated with Ship Movements between Ports in the European Community*, Table 2.10, prepared for the European Commission, July 2002.

[61] California Air Resources Board, *2005 Oceangoing Ship Survey, Summary of Results,* September 2005.

Table A-27. Tier 0 Auxiliary Engine Emission Factors by Ship Type, g/kWh

Ship Type	Fuel Sulfur	Emission Factors (g/kWh)							
		NOx	PM$_{10}$	PM$_{2.5}$	HC	CO	SO$_2$	CO$_2$	BSFC
Passenger	2.56%	14.6	1.36	1.25	0.4	1.1	11.36	718	226
	0.10%	13.9	0.18	0.17	0.4	1.1	0.42	691	217
Other	2.21%	14.5	1.16	1.07	0.4	1.1	9.74	707	224
	0.10%	13.9	0.18	0.17	0.4	1.1	0.42	691	217

A.6.6.3. Fuel Sulfur Levels

Where RO is used in propulsion or auxiliary engines, it is assumed to be 2.7% sulfur for all ports for the 2011 baseline analysis. In the C3 RIA, West Coast ports were assumed to use 2.5% sulfur RO instead of 2.7% sulfur. Using 2.7% sulfur RO in Washington and Oregon ports is consistent with the 2011 Puget Sound inventory[62] where 2.7% sulfur RO was assumed.

A.6.6.4. Black Carbon

BC is the light-absorbing component of particulate matter and is formed by the incomplete combustion of carbon-based fuels. Like CO$_2$, BC is a global warming pollutant. EPA's Report to Congress on Black Carbon[63] lists 0.03 as the BC/PM$_{2.5}$ factor to use for C3 commercial marine vessels for all engine types and fuels for 2011.

A.6.6.5. Treatment of Electric-Drive Ships

Many passenger ships and tankers have either diesel-electric or gas turbine-electric engines that are used for both propulsion and auxiliary purposes. Lloyd's identifies these types of engines in its database and that information was used to distinguish them from direct and geared drive systems for this analysis. Generally, the power Lloyd's lists is the total power for electric drive vessels. To separate out propulsion from auxiliary power for purposes of calculating emissions, the total power listed in the Lloyd's data was divided by one plus the ratio of auxiliary to propulsion power given in Table A-20 to give the propulsion power portion and the remaining portion was considered auxiliary engine power.[64] In addition, no low load adjustment factor was applied to diesel electric engines for loads below 20% because several engines are used to generate power, and some can be shut down to allow others to operate at a more efficient setting. This same methodology was used for the calculations in the C3 RIA.

---

[62] Starcrest Consulting Group, *2011 Puget Sound Maritime Emission Inventory,* September 2012.
http://www.pugetsoundmaritimeairforum.org/uploads/PV_FINAL_POT_2011_PSEI_Report_Update_23_May_13_scg.pdf
[63] U.S. Environmental Protection Agency, *Report to Congress on Black Carbon,* EPA-450/R-12-001, March 2012, p. 87.
[64] ICF International, *Commercial Marine Port Development – 2002 and 2005 Inventories,* September 2007.

A.6.7. Bin Mismatches

In some cases, the ship type/engine type/DWT range bin in the modeled port was not provided in the typical port inventory. In that case, the nearest match to the bin at the modeled port was used from the typical port inventory. This same methodology of using near bins was used in development of the C3 RIA. Table A-28 shows the typical port bins used when there was no exact match for the bin at the modeled port. While the same methodology was used, it is possible that the bin matched for the C3 RIA might be different to the bin matched in this analysis.

Table A-28. Typical Port Bin Mismatches

Port	Modeled Port Bin			Typical Port Bin		
	Ship Type	Engine Type	DWT Range	Ship Type	Engine Type	DWT Range
Port of New Orleans	CONTAINER SHIP	SSD	> 90,000	CONTAINER SHIP	SSD	45,000 – 90,000
	PASSENGER	MSD-ED	10,000 – 20,000	PASSENGER	MSD-ED	< 10,000
	TANKER	MSD-ED	< 30,000	TANKER	MSD	< 30,000
			30,000 – 60,000			30,000 – 60,000
Port of Miami	CONTAINER SHIP	SSD	35,000 – 45,000	CONTAINER SHIP	SSD	25,000 – 35,000
			45,000 – 90,000			
	PASSENGER	MSD-ED	< 10,000	PASSENGER	MSD	< 10,000
			10,000 – 20,000			
		GT-ED	10,000 – 20,000			
Port of Mobile	BULK CARRIER	MSD	45,000 – 90,000	BULK CARRIER	SSD	45,000 – 90,000
			> 90,000			> 90,000
	CONTAINER SHIP	SSD	35,000 – 45,000	CONTAINER SHIP	SSD	25,000 – 35,000
			45,000 – 90,000		MSD	45,000 – 90,000
	MISCELLANEOUS	MSD	All	MISCELLANEOUS	SSD	All
		MSD-ED	All			
Port of Baltimore	AUTO CARRIER	SSD	> 30,000	AUTO CARRIER	SSD	20,000 – 30,000
	CONTAINER SHIP	SSD	> 90,000	CONTAINER SHIP	SSD	45,000 – 90,000
	PASSENGER	SSD	< 10,000	PASSENGER	ST	< 10,000
	RORO	MSD	10,000 – 20,000	RORO	MSD	20,000 – 30,000
Port of Norfolk (Hampton Roads)	CONTAINER SHIP	SSD	> 90,000	CONTAINER SHIP	SSD	45,000 – 90,000
	PASSENGER	MSD-ED	10,000 – 20,000	PASSENGER	MSD-ED	< 10,000
		SSD	< 10,000		ST	< 10,000
	TANKER	SSD	120,000 – 150,000	TANKER	SSD	> 150,000
Port of Philadelphia	AUTO CARRIER	SSD	> 30,000	AUTO CARRIER	SSD	20,000 – 30,000
	CONTAINER SHIP	SSD	35,000 – 45,000	CONTAINER SHIP	SSD	25,000 – 35,000
			45,000 – 90,000			
	RORO	MSD	10,000 – 20,000	RORO	MSD	< 10,000
		SSD	20,000 – 30,000		SSD	10,000 – 20,000
Port of Corpus Christi	AUTO CARRIER	SSD	20,000 – 30,000	AUTO CARRIER	SSD	10,000 – 20,000
			> 30,000			
	BULK CARRIER	MSD-ED	45,000 – 90,000	BULK CARRIER	SSD	45,000 – 90,000
	MISCELLANEOUS	MSD-ED	All	MISCELLANOUS	MSD	All
	TANKER	ST	60,000 – 90,000	TANKER	ST	30,000 – 60,000
	TUG	MSD	All	MISCELLANEOUS	MSD	All

Port	Modeled Port Bin			Typical Port Bin		
	Ship Type	Engine Type	DWT Range	Ship Type	Engine Type	DWT Range
Port Tampa Bay	CONTAINER SHIP	SSD	45,000 – 90,000	CONTAINER SHIP	MSD	45,000 – 90,000
	MISCELLANEOUS	MSD-ED	All	MISCELLANEOUS	SSD	All
	PASSENGER	MSD-ED	< 10,000	PASSENGER	MSD	< 10,000
		GT-ED	10,000 – 20,000			
	REEFER	SSD	20,000 – 30,000	REEFER	SSD	10,000 – 20,000
	RORO	MSD	10,000 – 20,000	RORO	SSD	10,000 – 20,000
Port of Savannah	AUTO CARRIER	SSD	> 30,000	AUTO CARRIER	SSD	20,000 – 30,000
	CONTAINER SHIP	SSD	> 90,000	CONTAINER SHIP	SSD	45,000 – 90,000
	TANKER	MSD-ED	60,000 – 90,000	TANKER	MSD	30,000 – 60,000
		SSD	120,000 – 150,000		SSD	> 150,000
		ST	60,000 – 90,000		ST	30,000 – 60,000
			90,000 – 120,000			
Port of San Juan	CONTAINER SHIP	SSD	35,000 – 45,000	CONTAINER SHIP	SSD	25,000 – 35,000
		SSD	45,000 – 90,000		MSD	45,000 – 90,000
		ST	25,000 – 35,000		SSD	25,000 – 35,000
	MISCELLANEOUS	ST	All	MISCELLANEOUS	SSD	All
	PASSENGER	MSD-ED	10,000 – 20,000	PASSENGER	MSD	< 10,000
		SSD	< 10,000		ST	< 10,000
		GT-ED	10,000 – 20,000		ST	10,000 – 20,000

A.6.8. Matching Lloyd's Data to USACE Entrances and Clearance Data

To match activity data with the correct emission factors, the Entrances and Clearances data were matched to Lloyd's Register of Ships. The Entrances and Clearances data[65] contain the following information for each major port or waterway:

- date a vessel made entry into (entrance record) or cleared (clearance record) the U.S. Customs port;

- vessel's full name;

- type of vessel by one digit rig type or International Classification of Ships by Type (ICST) code;

- vessel's flag of registry;

- vessel's previous (entrance record) or next (clearance record) port of call, whether the port was foreign or domestic;

- vessel's Net and Gross Registered Tonnage;

- vessel's draft (feet); and

- vessel's International Maritime Organization number.

---

[65] Available at http://www.navigationdatacenter.us/data/dataclen.htm.

Since this does not contain any call time-in-mode information, average time in mode and speeds need to be used with the USACE data to estimate emissions at ports (see section A.6.4).

It is important to note that these data only represent foreign cargo movements, and does not account for U.S. ships delivering domestic cargo from one U.S. port to another U.S. port (covered under the Jones Act[66]). However, U.S. flagged ships carrying foreign cargo from a foreign port to a U.S. port or from a U.S. port to a foreign port are accounted for in the data as these are considered foreign cargo movements. At most ports, domestic commerce is carried out by Category 2 ships, although there are a few exceptions, particularly on the West Coast. A study by ICF found that the USACE Entrances and Clearances data accounted for more than 90% of the emissions from Category 3 ships calling on U.S. ports, so neglecting Jones Act ships is assumed to be small.[67]

Additionally, the Entrances and Clearances data do not always match MEPA data because Entrances and Clearances include cargo movements from both public and private terminals at a port while the MEPA data usually only cover calls at public terminals, as Port Authorities generally do not have jurisdiction over private terminals.

Entrances and Clearances data for 2011 contained over 100,000 individual entrances or clearances by ships, tugs, and barges for U.S. ports or waterways. The ports of interest were matched to Entrances and Clearances data as shown in Table A-29.

Table A-29. Corresponding USACE Port Names

Port Name	USACE Port #	USACE Name
Port of New York and New Jersey	398	Consolidated Port of New York
Port of New Orleans	2251	Port of New Orleans, LA
Port of Miami	2164	Miami Harbor, FL
Port of South Louisiana	2253	Port of South Louisiana (LA)
Port of Seattle	4722	Seattle Harbor, WA
Port of Baton Rouge	2252	Port of Baton Rouge, LA
Port Arthur	2416	Port Arthur, TX
Port of Portland	4644	Port of Portland, OR
Port of Mobile	2005	Mobile Harbor, AL
Port of Houston	2012	Houston Ship Channel, TX (Houston, TX)
Port of Baltimore	700	Baltimore Harbor and Channels, MD
Port of Hampton Roads (Norfolk)	744	Elizabeth River, VA
Port of Philadelphia	552	Philadelphia Harbor, PA
Port of Charleston	773	Charleston Harbor, SC
Port of Corpus Christi	2414	Corpus Christi, TX
Port Tampa Bay	2021	Tampa Harbor, FL
Port of Savannah	776	Savannah Harbor, GA
Port of Coos Bay, OR	4660	Coos Bay, OR
Port of San Juan, PR	2136	San Juan Harbor, PR

---

[66] Merchant Seaman Protection and Relief 46 USCS Appx § 688 (2002) Title 46. Appendix. Shipping Chapter 18.·
[67] ICF International, *Inventory Contribution of U.S. Flagged Vessels*, June 2008.

As done in the C3 RIA, barges that are not self-propelled were removed from the data. These are indicated by the Rig field, where Rig = 4 indicates a dry barge while Rig = 5 represents a liquid barge. All barges that were not part of an integrated tug-barge (ITB) were eliminated.

After eliminating barges and calls at ports outside the scope of this assessment, there were over 77,000 records for 7,696 vessels. Of these vessels, almost 99% could be matched to Lloyd's data using IMO number and ship name. 18 vessels had incorrect IMO numbers, 15 of which were matched by IMO number and gross tons or by vessel name and gross tons. 86 vessels did not have IMO numbers; 21 of these were matched by vessel name and gross tons or by vessel name and ship type. This left 68 unmatched vessels, of which 60 were known to be less than 100 gross tons.

Of the 7,696 unique vessels, 93% were determined to be Category 3 by calculating cylinder displacement from bore size and stroke length. 0.2% were steam turbine driven and 0.1% were gas turbine driven.[68] Of the Category 3 engines, gas turbines, and steam turbines, 6,459 were slow speed diesels (SSD) 2-stroke engines, 607 were medium speed diesels (MSD) 4-stroke engines, 97 were medium speed diesels with electric drive (MSD-ED), nine were gas turbine electric drive (GT-ED), and 16 were steam turbine driven (ST).

After determining the engine type for each vessel in the Entrances and Clearances data, the vessels were assigned a ship type based upon Lloyd's categorization as shown in Table A-30.

Table A-30. Corresponding Ship Types

Ship Type	Lloyd's Ship Type
Auto Carrier	Vehicles Carrier
Bulk Carrier	Barge Carrier[69]
	Bulk Carrier
	Bulk Carrier (with Vehicle Decks)
	Bulk Carrier, Laker Only
	Bulk Carrier, Self-discharging
	Cement Carrier
	Fish Carrier
	Fish Factory Ship
	Heavy Load Carrier
	Heavy Load Carrier, semi-submersible
	Livestock Carrier
	Ore Carrier
	Rail Vehicles Carrier
	Wood Chips Carrier
Container Ship	Container Ship (Fully Cellular with RORO Facility)
	Container Ship (Fully Cellular)
	Container/RORO Cargo Ship

---

[68] While steam and gas turbines are not diesel engines and thus do not fall into normal Categories 1, 2, or 3, they were included in the analysis as OGV propulsion engines.

[69] Barge carriers were originally separated in the C3 RIA but are such a small category that they are combined here with bulk carriers.

Ship Type	Lloyd's Ship Type
General Cargo	General Cargo Ship
	General Cargo Ship (with RORO facility)
	Open Hatch Cargo Ship
	Palletized Cargo Ship
Miscellaneous	Anchor Handling Tug Supply
	Offshore Support Vessel
	Pipe Layer
	Pipe Layer Crane Vessel
	Research Survey Vessel
	Trailing Suction Hopper Dredger
	Training Ship
	Trawler
	Yacht Carrier, semi-submersible
Passenger	Passenger/Cruise
	Passenger/RORO Ship (Vehicles)
Reefer	Fruit Juice Carrier, Refrigerated
	Refrigerated Cargo Ship
RORO	Logistics Vessel (Naval RORO Cargo)
	RORO Cargo Ship
Tanker	Asphalt/Bitumen Tanker
	Bulk/Oil Carrier (OBO)
	Chemical Tanker
	Chemical/Products Tanker
	Combination Gas Tanker (LNG/LPG)
	Crude Oil Tanker
	Crude/Oil Products Tanker
	FPSO, Oil
	LNG Tanker
	LPG Tanker
	LPG/Chemical Tanker
	Molten Sulfur Tanker
	Products Tanker
	Shuttle Tanker
Tug	Articulated Pusher Tug
	Tug

The final step in the matching process was to fill data gaps. For example, three ships did not have service speeds. The average value for the Lloyd's ship type was used for those three vessels as was done in the C3 RIA. Finally, DWT ranges were assigned to the various vessels as was done in 2002 in the C3 RIA.

A.6.9. Additional Details for RSZ, Maneuvering, and Hoteling Calculations

The Port of Houston has two reduced speed zones to accommodate different speeds within the Houston Ship Channel. These are shown in Figure A-3. The first is in Galveston Bay starting at Barbours Cut (see red line). The longer distance is for those ships coming from the ship channel past Barbours Cut while the shorter distance is for those ships leaving Bayport. The second is from where the Houston Ship

Channel widens around Channelview to Barbours' Cut (see green line). These two RSZs were combined in the C3 RIA but were separated out in this analysis to provide more detail.

Figure A-3. Houston Reduced Speed Zones[70]

It is assumed that all ships except for container ships and passenger ships enter the Houston ship channel and head toward the Turning Basin. They travel over both RSZ 1 and RSZ 2 as well as maneuver to and from the Turning Basin. Passenger ships are assumed to all stop at Bayport, so they only travel down a portion of RSZ 1 and then maneuver into Bayport. Most container ships stop at either Bayport or Barbours Cut but some smaller container ships head toward the turning basin. Those that stop at Barbours Cut travel all the way down RSZ 1 but only 0.7 km down RSZ 2 and then maneuver into Barbours Cut. The distribution of container ships that stop at Bayport, Barbours Cut, and the Turning Basin are shown in Table A-31. These were determined from the Entrances and Clearances data and Lloyd's data by examining the operating company for each ship and assigning it to one of the stops based upon the Port of Houston website.[71]

Table A-31. Distribution of Container Ship Stops at Port of Houston

DWT	Barbours Cut	Bayport	Turning Basin
< 25,000	4.3%	65.7%	30.1%
25,000 – 35,000	55.6%	44.4%	0.0%
35,000 – 45,000	60.6%	39.4%	0.0%
45,000 – 90,000	24.9%	75.1%	0.0%
> 90,000	0.0%	100.0%	0.0%

---

[70] Map data: Google.

[71] Available at: http://www.portofhouston.com/container-terminals.

Maneuvering and hoteling times for the Ports of Seattle and New York/New Jersey came from updated port inventories that did not provide these data for the same bins used for the other ports.[72] This information was calculated for the Ports of Seattle and New York/New Jersey from movement data.

The movement data taken from the two inventories are shown in Table A-32.

Table A-32. Arrivals, Departures, and Shifts for Ports of Seattle and New York/New Jersey from Published Inventories

Port	Ship Type	Bin	Arrivals	Departures	Shifts
Seattle	Bulk	All	2	76	105
	Container	1000 TEU	108	116	15
		2000 TEU	73	89	24
		3000 TEU	29	30	1
		4000 TEU	111	115	6
		5000 TEU	124	126	5
		6000 TEU	17	68	51
		7000 TEU	83	81	1
		8000 TEU	130	135	8
		9000 TEU	2	2	0
		10000 TEU	1	1	0
	General Cargo	All	14	39	26
	Passenger	All	196	196	0
	Tanker	All	0	5	0
	Tug	All	0	1	4
New York/New Jersey	Auto Carrier	All	266	265	93
	Bulk Carrier	All	59	58	75
	Container	All	2,033	2,032	71
	General Cargo	All	30	29	8
	Passenger	All	97	97	0
	Reefer	All	46	46	2
	RORO	All	90	91	55
	Tanker	All	76	76	126

To calculate maneuvering time at the two ports, calls were defined as the maximum of published arrivals and departures as used in this analysis. Estimated port entrance to berth distances (one-way maneuvering distance) and shift distances were estimated using Google Earth. Berthing times were estimated at 0.5 hours. These assumptions are shown in Table A-33 for the two ports. An average maneuvering speed of 4 knots was used in the calculations to simulate the stop/start nature of

---

[72] Maneuvering times for Port of Charleston also came from an updated inventory; however, these were listed in the published inventory document so no calculations for maneuvering time needed to be done.

maneuvering in a port area. The maneuvering time calculated just includes maneuvering movements. No anchorage time is included because it is not provided in the two inventories.

Table A-33. Estimated Distances and Times for Maneuvering Calculations

Port	Item	Value
Port of Seattle	One way maneuvering distance	1.8 nm
	Berthing time	0.5 hrs per call
	Shifting Distance	1.75 nm
Port of New York and New Jersey	One way maneuvering distance	4.0 nm
	Berthing time	0.5 hr per call
	Shifting Distance	7.25 nm

The information in Table A-32 is used together with the published inventory arrival, departure, and shift information in Table A-33 to calculate maneuvering times using the equation below:

$$\text{Maneuvering Time (hr/call)} = ((\text{Arr}+\text{Dep})*\text{OWMD}/4 + \text{Calls}*\text{BT}+\text{S}*\text{SD}/4)/\text{Calls} \qquad \text{Eq. A-8}$$

Where

> Arr = number of arrivals,
> Dep = number of departures,
> OWMD = one-way maneuvering distance in nm,
> BT = berthing time in hours,
> S = number of shifts,
> SD = average shift distance in nm,
> Calls = maximum of arrivals and departures, and
> 4 = average maneuvering speed in knots.

Both the one-way maneuvering distance multiplied by the sum of arrivals and departures and the number of shifts times the average shift distance are divided by the average maneuvering speed of 4 knots to obtain hours of maneuvering. Berthing time in hours per call is then multiplied by calls and added to the maneuvering total. The total is then divided by the number of calls to determine maneuvering time per call in hours. Calculated maneuvering times are shown in Table A-34 for the two ports.

Table A-34. Calculated Maneuvering Times for Port of Seattle

Port	Ship Type	Bin	Maneuvering Time (hrs)
Port of Seattle	Bulk	All	1.57
	Container	1000 TEU	1.43
		2000 TEU	1.44
		3000 TEU	1.40
		4000 TEU	1.41
		5000 TEU	1.41
		6000 TEU	1.39
		7000 TEU	1.39
		8000 TEU	1.41
		9000 TEU	1.40
		10000 TEU	1.40
	General Cargo	All	1.40
	Passenger	All	1.40
	Tanker	All	0.95
	Tug	All	2.70
Port of New York/New Jersey	Auto Carrier	All	1.55
	Bulk Carrier	All	1.95
	Container	All	1.42
	General Cargo	All	1.50
	Passenger	All	1.40
	Reefer	All	1.42
	RORO	All	1.66
	Tanker	All	2.13

Maneuvering times for container ships for the Port of Seattle had to be translated from the TEU bins listed in the published inventory to the DWT bins used in this analysis. To accomplish this, container ships at Port of Seattle in the Entrances and Clearances data were defined both ways and weighting factors determined for each category. These weighting factors are shown in Table A-35. For example, to calculate maneuvering and hoteling time for the 25,000–35,000 DWT bin, maneuvering and hoteling times for 1000 TEU and 2000 TEU container ships were weighted by 35.8% and 64.2%, respectively.

Table A-35. Container Ship Size Translations for Port of Seattle

DWT Range	TEUs	Percent
< 25,000	1000	100.0%
25,000 – 35,000	1000	35.8%
	2000	64.2%
35,000 – 45,000	2000	91.6%
	3000	8.4%
45,000 – 90,000	2000	0.0%
	3000	0.9%
	4000	33.9%
	5000	47.8%
	6000	17.3%
	7000	0.0%
> 90,000	6000	1.1%
	7000	37.1%
	8000	59.6%
	9000	1.8%
	10000	0.4%

The result of weighting the maneuvering times shown in Table A-34 by the weights in Table A-35 is shown in Table A-36. There was no container ship detail on maneuvering times (by TEU) in the New York/New Jersey inventory.

Table A-36. Translated Container Ship Maneuvering Times for Port of Seattle

DWT Bin	Maneuver Time (hr)
< 25,000	1.43
25,000 – 35,000	1.43
35,000 – 45,000	1.43
45,000 – 90,000	1.40
> 90,000	1.40

Hoteling time was also taken from the Ports of Seattle, New York/New Jersey, and Charleston published inventories and translated into the DWT bins used in this analysis. Hoteling times for Port of Seattle and Port of New York/New Jersey are provided in Table A-37.

Table A-37. Published Hoteling Times for the Ports of Seattle and New York/New Jersey

Ship Type	Bin	Average Hoteling Time (hrs)	
		Seattle	New York
Auto Carrier	All	ni [a]	15
Bulk	All	88.0	35
Container	1000 TEU	24.2	18
	2000 TEU	30.3	16
	3000 TEU	31.8	22
	4000 TEU	31.6	20
	5000 TEU	30.0	24
	6000 TEU	28.7	30
	7000 TEU	27.8	36 [b]
	8000 TEU	38.3	41
	9000 TEU	33.3	40
	10000 TEU	32.1	ni
General Cargo	All	30.1	14
Passenger	All	10.1	10
Tanker	Chemical	N/A	29
	HandySize	ni	2
	PanaMax	ni	5
Tug	All	110.2	ni

[a] ni = not included; N/A = not available

[b] Value not listed for 7000 TEU. Taken as the average of 6000 TEU and 8000 TEU.

The Seattle inventory shows a value of "N/A" for tanker ship hoteling time. Since tankers represent a small portion of the Seattle inventory, no tanker hoteling emissions were calculated for Seattle in this analysis.

Hoteling times for container ships for both ports and tanker ships for New York/New Jersey were translated to the DWT bins used in this analysis. To accomplish this for the Port of New York/New Jersey, container and tanker ships in the Entrances and Clearances data for that port were processed as described above for Seattle. The results are given below in Table A-38.

Table A-38. Container and Tanker Ship Size Translations for Port of New York/New Jersey

Ship Type	DWT Range	TEUs	Percent
Container Ship	< 25,000	1000	100.0%
	25,000 – 35,000	1000	35.8%
		2000	64.2%
	35,000 – 45,000	2000	91.6%
		3000	8.4%
	45,000 – 90,000	3000	0.9%
		4000	33.9%
		5000	47.8%
		6000	17.3%
	> 90,000	6000	1.1%
		7000	37.1%
		8000	59.6%
		9000	1.8%
		10000	0.4%
Tankers	< 30,000	Chemical	8.3%
		HandySize	91.7%
	30,000 – 60,000	Chemical	0.2%
		HandySize	99.8%
	> 60,000	PanaMax	100.0%

The container hoteling times for Port of Seattle translated into the DWT range bins used in this analysis are shown in Table A-39. The translated container and tanker ship hoteling times for Port of New York/New Jersey are shown in Table A-40.

Table A-39. Translated Container Ship Hoteling Times for Port of Seattle

DWT Bin	Hoteling Time (hr)
< 25,000	24.2
25,000 – 35,000	28.1
35,000 – 45,000	30.4
45,000 – 90,000	30.3
> 90,000	34.2

Table A-40. Translated Container and Tanker Ship Hoteling Times for Port of New York/New Jersey

Ship Type	DWT Bin	Hoteling Time (hr)
Container Ship	< 25,000	18.0
	25,000 – 35,000	16.5
	35,000 – 45,000	17.8
	45,000 – 90,000	21.5
	> 90,000	40.5
Tankers	< 30,000	4.2
	30,000 – 60,000	2.1
	> 60,000	5.0

The Charleston Inventory included hoteling time by ships and calls. This was mapped to the ships and calls found in the Entrances and Clearances data for the Port of Charleston and averaged by ship type/engine type/DWT bin. The results from this analysis are shown in Table A-41.

Table A-41. Calculated Hoteling Times for Port of Charleston

Ship Type	Main Engine	DWT Range	Hoteling Time (hr)
Auto Carrier	SSD	10,000 – 20,000	38.0
		20,000 – 30,000	38.0
		> 30,000	38.0
Bulk Carrier	MSD	< 25,000	20.3
	SSD	< 25,000	20.3
		25,000 – 35,000	20.3
		35,000 – 45,000	13.0
		45,000 – 90,000	28.2
Container Ship	MSD	< 25,000	11.1
	SSD	< 25,000	13.8
		25,000 – 35,000	14.1
		35,000 – 45,000	14.9
		45,000 – 90,000	14.7
		> 90,000	29.9
General Cargo	MSD	< 25,000	26.8
	SSD	< 25,000	36.3
		25,000 – 35,000	30.3
		35,000 – 45,000	22.8
		45,000 – 90,000	21.5
Passenger	MSD	< 10,000	46.2
	MSD-ED	< 10,000	46.2
		10,000 – 20,000	46.2
	SSD	< 10,000	46.2
Reefer	SSD	< 10,000	62.5
RORO	MSD	< 10,000	26.8
	SSD	10,000 – 20,000	13.3
		> 30,000	13.3
	GT-ED	> 30,000	17.5
Tanker	MSD	< 30,000	63.7
	SSD	< 30,000	54.4
		30,000 – 60,000	66.2
		60,000 – 90,000	73.5
		90,000 – 120,000	79.6

Appendix B. Business as Usual Emission Inventory Methodology

B.1. Introduction

This assessment included the development of representative, national scale inventories for the baseline and Business as Usual (BAU) cases for different pollutants and years, followed by the analysis of various strategies to reduce port-related mobile source emissions. This appendix details the methodology used to develop the BAU emission inventories for the calendar years 2020, 2030, and 2050.

Separate inventories for various pollutants were developed for the drayage trucks, rail, cargo handling equipment (CHE), harbor craft, and ocean going vessels (OGV) sectors. The following pollutants were included in these inventories for 2020 and 2030: nitrogen oxides (NOx), fine particulate matter (PM$_{2.5}$), volatile organic compounds (VOCs), sulfur dioxide (SO$_2$), carbon dioxide (CO$_2$), black carbon (BC), acetaldehyde, benzene, and formaldehyde. Note that the selected air toxics (acetaldehyde, benzene, and formaldehyde) were only analyzed for the non-OGV sectors and SO$_2$ was only analyzed for the OGV sector. Additionally, inventories were developed for 2050 for CO$_2$ only. In general, inventories were developed for each port analyzed in this assessment using national scale methodology and data, although port-specific data and adjustments were included where available and are noted where appropriate in this appendix. This assessment does not provide specific projections for local decision-making at individual ports or specific neighborhoods.

B.2. Projecting the Baseline Inventory

The baseline inventory was projected using port and sector specific growth factors. Then, adjustments were made at some ports due to recent or planned changes that are expected to change future emissions.

B.2.1. Growth Factors

For the OGV sector, growth was based on regional annual bunker fuel growth rates from 2002 to 2020 in a 2008 study by Research Triangle Institute.[73] Average annual growth factors by region that were derived from that study are listed in the C3 RIA Table 3-69. A subset of those, listed in Table B-1, were used for OGV sector growth for both 2020 and 2030.

Growth for other sectors was based on international trade growth factors from the same study. Compound annual growth rates (CAGR) for years 2020 and 2030 were calculated from commodity movements, which are imports plus exports. CAGR were determined relative to the 2011 baseline year

---

[73] Research Triangle Institute, *Global Trade and Fuel Assessment – Future Trends and Effects of Requiring Cleaner Fuels in the Marine Sector*, EPA Report EPA420-R-08-021, November 2008.

and then aggregated into the four conveyance methods shown in Table B-2. The 2011 US Army Corps of Engineers (USACE) commodity throughput[74] at each port of interest was used to weight the various categories in Table B-2 to determine the 2020 and 2030 CAGR for each port. For example, if the 2011 commodity throughput at Port X on the Atlantic Coast was 50% containers, 35% bulk, and 15% liquid, the 2020 CAGR for Port X would be calculated as:

$$\left(\frac{50\% \times 4\% \ + \ 35\% \times 3.2\% \ + \ 15\% \times 0.5\%}{50\% + 35\% + 15\%}\right) = 3.2\%$$

Table B-1. Regional Bunker Fuel Use Growth Factors for 2020 and 2030

Region	Average Annual Growth Rate
East Coast	4.5%
Gulf Coast	2.9%
South Pacific	5.0%
North Pacific	3.3%

Table B-2. Compound Annual Growth Rates for 2020 and 2030 by Region and Commodity

Conveyance Category	U.S. ATLANTIC – Imports + Exports		U.S. PACIFIC NORTH – Imports + Exports		U.S. PACIFIC SOUTH – Imports + Exports		U.S. GULF COAST – Imports + Exports	
	2020	2030	2020	2030	2020	2030	2020	2030
Bulk	3.2%	2.7%	4.0%	4.0%	3.9%	3.8%	3.3%	3.2%
Container	4.0%	4.4%	4.0%	4.5%	4.3%	4.9%	3.8%	4.1%
Liquid	0.5%	1.1%	1.5%	1.6%	1.1%	1.1%	1.4%	1.6%
Other	5.0%	4.9%	5.0%	4.8%	7.4%	7.2%	3.9%	4.2%
Total	**2.7%**	**2.9%**	**3.8%**	**4.0%**	**3.5%**	**4.0%**	**2.2%**	**2.3%**

B.2.2. Adjustments to the Projected BAU

In addition to baseline growth, the projected BAU inventory also considered ongoing or planned changes in port operations that could substantially change emissions in future years. An example of such a change could be plans for construction of on-dock rail that would change the mode split and shift cargo from truck to rail. Where identified, options were assessed for quantifying these changes in the BAU emission projections.

B.3. Drayage Trucks

This section describes the methodology used for projecting BAU emissions from drayage activity.

---

[74] Available at: http://www.navigationdatacenter.us/db/wcsc/archive/xls/man11/.

B.3.1. Modeling Approach

Truck activity in 2020 and 2030 was estimated by applying the growth factors described in Table B-2 to the 2011 baseline cargo volumes. The share of cargo throughput moved by truck came from version 3 of the Freight Analysis Framework (FAF).[75] The FAF modal splits by commodity class were aggregated to total tonnage share moved by truck. The same modal splits were used for 2011, 2020, and 2030.

DrayFLEET[76] was run with the baseline 2011 cargo volumes and the national 2020 age distribution from MOVES2010b.[77] These intermediate emission inventories were then scaled by the ratio of projected truck tonnage in 2020 to the baseline 2011 truck tonnage to calculate the 2020 drayage BAU inventory. This was then repeated with the national 2030 age distribution from MOVES2010b and projected 2030 tonnage. However, since the version of DrayFLEET used in this analysis was not capable of running calendar year 2030, the model was run as it was for 2020, but with a modified 2030 age distribution. To get the model to accept the 2030 age distribution, all 2021-2030 model year vehicles were labeled as 2020. This was not expected to impact the results, as the heavy duty vehicle standards modeled here are identical for model years 2020-2030.[78] To calculate the 2050 CO_2 inventories, the 2030 inventories were scaled using the anticipated growth in tonnage.

B.3.2. Additional Pollutants

The DrayFLEET Model estimates emissions for $PM_{2.5}$, NOx, HC, CO, and CO_2. VOC emissions were estimated as equal to the HC emissions. Emissions for the air toxics formaldehyde, benzene, and acetaldehyde were estimated as a fraction of VOC emissions based on diesel speciation profiles calculated from running MOVES2010b[79], as shown in Table B-3. These fractions vary depending on whether or not VOC is controlled (model year 2007 and later). Weighted speciation factors were calculated for 2020 and 2030 based on the percent of 2007 model year and greater trucks in the fleet. Approximately 50% of drayage trucks in 2020 and 11% in 2030 were model year 2006 or earlier.

Black carbon (BC) emissions were estimated as a 77% of $PM_{2.5}$ emissions, consistent with EPA's Report to Congress on Black Carbon.[80]

---

[75] Available at: http://www.ops.fhwa.dot.gov/freight/freight_analysis/faf/.

[76] U.S. Environmental Protection Agency, *SmartWay DrayFLEET: Truck Drayage Environment and Energy Model, Version 2.0 User's Guide*, EPA-420-B-12-065, June 2012.

[77] U.S. Environmental Protection Agency, *Motor Vehicle Emission Simulator (MOVES): User Guide for MOVES2010b*, EPA-420-B-12-001b, June 2012.

[78] This modeling relies on a previous version of MOVES that does not include EPA's heavy-duty engine and vehicle GHG regulations.

[79] U.S. Environmental Protection Agency, *MOVES2010b: Additional Toxics Added to MOVES*. EPA-420-B-12-029a, May 2012, Sec 3.1.1. Available at: http://www.epa.gov/otaq/models/moves/documents/420b12029a.pdf.

[80] U.S. Environmental Protection Agency, *Report to Congress on Black Carbon*, EPA-450/R-12-001, March 2012, p. 87.

Table B-3. Diesel Truck Air Toxic Speciation Profiles Based on MOVES2010b

Pollutant	Toxic/VOC without Control	Toxic/VOC with Control
Acetaldehyde	0.035559	0.06934
Benzene	0.007835	0.01291
Formaldehyde	0.078225	0.21744

B.4. Rail

This section describes the methodology used for projecting BAU emissions from port-related rail activity.

B.4.1. Modeling Approach

The baseline 2011 rail sector activity was grown to 2020 and 2030 using the commodity growth rates shown in Table B-2. Emission factors were calculated from the baseline inventories, adjusted for expected changes in the future fleet, and then applied to the projected activity to determine the BAU inventories.

Gross emission factors[81] were calculated from the baseline 2011 rail inventory using the following equation:

$$EF_{2011} = E_{2011}/(C_{2011} \times S)$$

Eq. B-1

Where:

EF = Emission factor for a specific pollutant, port, and locomotive type (g/ton),

E = Total annual emissions for a specific pollutant, port, and locomotive type (g),

C = Total cargo throughput for a specific port (tons), and

S = Share of cargo throughput moved by rail for a specific port (percent of total cargo tonnage).

To calculate the gross emission factors, the total annual emissions came from the baseline rail inventories (see Appendix A). The total cargo throughput came from USACE's Waterborne Commerce Statistics.[82] The share of cargo throughput moved by rail came from version 3 of the Freight Analysis Framework (FAF).[83] The FAF modal splits by commodity class were aggregated to total tonnage share moved by rail. The same modal splits were used for 2011, 2020, and 2030. Combining all of these results in gross emission factors that are valid for the 2011 locomotive fleet at each port.

However, since fleets turn over to newer models in future years that meet stricter emission standards, the gross emission factors were scaled for use in 2020 and 2030 based on projected future fleet

---

[81] Used here, a "gross emission factor" is an estimate of emissions per unit goods moved. It is described as "gross" to distinguish from more refined factors, such as engine-, equipment-, operations-, or technology-specific emission factors determined from sources such as engine certification or emissions models.

[82] Available at: http://www.navigationdatacenter.us/db/wcsc/archive/xls/man11/.

[83] Available at: http://www.ops.fhwa.dot.gov/freight/freight_analysis/faf/.

emission factors listed in EPA's 2008 Locomotive and Marine Emission Standards Rulemaking.[84] The emission factors given in the rulemaking were not used directly because they are in terms of grams per gallon, and the units required here were grams per ton of cargo moved. However, their values as used in this scaling analysis are presented in Table B-4. Please note that only NOx, PM_{10}, and HC were available for this technique; the other pollutants were calculated separately. This scaling approach is consistent with EPA's 2011 Air Quality Modeling Platform's Technical Support Document.[85] Therefore, future year gross emission factors were calculated using the following equations:

$$EF_{2020} = EF_{2011} \times \left[\frac{FuelEF_{2020}}{FuelEF_{2011}}\right] \qquad \text{Eq. B-2}$$

$$EF_{2030} = EF_{2011} \times \left[\frac{FuelEF_{2030}}{FuelEF_{2011}}\right] \qquad \text{Eq. B-3}$$

Where:

EF = Emission factor for a specific pollutant, port, and locomotive type (g/ton) and

FuelEF = Fuel based emission factor by pollutant and locomotive type (Table B-4).

Table B-4. Projected Emission Factors (g/gal) from 2008 EPA Locomotive and Marine Emission Standards Rulemaking

Calendar Year	Pollutant	Large Line-haul	Large Switch
2011	NOx	149	235
	PM_{10}	4.4	5.3
	HC	7.7	14.0
2020	NOx	99	187
	PM_{10}	2.3	4.1
	HC	3.6	10.5
2030	NOx	53	119
	PM_{10}	1.0	2.5
	HC	1.9	6.2

Since the BAU inventories are grouped by rail lines and rail yards, it was assumed that baseline emissions associated with rail yards were from switcher locomotives, while all emissions associated with the rail line segments were from line-haul locomotives.

To calculate the BAU inventories, the future year emission factors were combined with grown rail activity as shown in the following equations:

---

[84] U.S. Environmental Protection Agency, *Emission Factors for Locomotives*, EPA-420-F-09-025, April 2009, Tables 5-7.

[85] U.S. Environmental Protection Agency, *Technical Support Document (TSD): Preparation of Emissions Inventories for the Version 6.0, 2011 Emissions Modeling Platform*, Section 4.4.1, February 26, 2014. Available at: http://www.epa.gov/ttn/chief/emch/2011v6/outreach/2011v6_2018base_EmisMod_TSD_26feb2014.pdf.

$$E_{2020} = EF_{2020} \times C_{2011} \times (1+G)^{(2020-2011)} \times S \qquad \text{Eq. B-4}$$

$$E_{2030} = EF_{2030} \times C_{2011} \times (1+G)^{(2030-2011)} \times S \qquad \text{Eq. B-5}$$

Where:

E = Total annual emissions for a specific pollutant, port, and locomotive type (g),

EF = Emission factor for a specific pollutant, port, and locomotive type (g/ton),

C = Total cargo throughput for a specific port (tons),

G = Commodity-based CAGR for the region in which the port is located (see Table B-2), and

S = Share of cargo throughput moved by rail for a specific port (percent of total cargo tonnage).

Separate emissions were calculated using the specific emission factors for line-hauls and switchers, both of which were associated with total cargo throughput handled by rail. Total rail emissions were then the sum of projected switcher plus line-haul emissions. Note that emission factors and activity data were used separately, rather than projection factors as used in the Mobile Source Air Toxics (MSAT) Rule.[86] National projection factors (including both growth and control) were not appropriate at the port level and would not accommodate the regional growth factors used here.

B.4.2. Additional Pollutants

The methodology described above provides inventories for NOx, PM_{10}, and HC only; therefore, VOC, $PM_{2.5}$, CO_2, BC, benzene, acetaldehyde, and formaldehyde needed be calculated separately.

B.4.2.1. VOC

VOCs were estimated to be 1.053 times HC emissions, which is consistent with EPA's 2008 Locomotive and Marine Emission Standards Rulemaking.

B.4.2.2. $PM_{2.5}$

$PM_{2.5}$ emissions were estimated to be 0.97 times the PM_{10} emissions, which is consistent with EPA's 2008 Locomotive and Marine Emission Standards Rulemaking.

B.4.2.3. BC

BC emissions were estimated to be 0.77 times the $PM_{2.5}$ emissions, which is consistent with EPA's Report to Congress on Black Carbon.[87]

---

[86] U.S. Environmental Protection Agency, *National Scale Modeling of Air Toxics for the Mobile Source Air Toxics Rule: Technical Support Document*, EPA 454/R-06-002, January 2006.

[87] U.S. Environmental Protection Agency, Report to Congress on Black Carbon, EPA-450/R-12-001, March 2012, p. 87.

B.4.2.4. CO₂

Future year CO_2 emission factors were assumed to remain constant and equal to the baseline year. For the 2050 CO_2 inventory, the 2030 CAGR were applied to grow the cargo throughput values to 2050 levels.

B.4.2.5. Air Toxics

Emissions for the air toxics formaldehyde, benzene, and acetaldehyde were estimated as a fraction of VOC emissions based on diesel speciation profiles calculated from running MOVES2010b. MOVES was used for the rail sector due to a lack of other sources more directly applicable to rail: NONROAD does not include locomotives and SPECIATE and the NEI do not include projections. The MOVES speciation profiles are shown in Table B-5. These fractions vary depending on whether or not VOC is controlled (model year 2007 and later). For nonroad engines, this relates primarily to engines mandated to use Tier 4 emission controls. Weighted speciation factors were calculated for 2020 and 2030 based on Tier 4 versus pre-Tier 4 engine distributions from the 2008 Locomotive and Marine RIA.[88] This is not consistent with the MSAT Rule or the Emissions Modeling Platform, which use constant speciation factors for future years.

Table B-5. Diesel Truck Air Toxic Speciation Profiles Based on MOVES2010b Applied to Rail[89]

Pollutant	Toxic/VOC without control	Toxic/VOC with Control
Acetaldehyde	0.035559	0.06934
Benzene	0.007835	0.01291
Formaldehyde	0.078225	0.21744

---

[88] U.S. Environmental Protection Agency, *Control of Emissions of Air Pollution from Locomotive Engines and Marine Compression Ignition Engines Less than 30 Liters Per Cylinder*, EPA420-R-08-001, March 2008.

[89] U.S. Environmental Protection Agency, *MOVES2010b: Additional Toxics Added to MOVES*. EPA-420-B-12-029a, May 2012, Sec 3.1.1. Available at: http://www.epa.gov/otaq/models/moves/documents/420b12029a.pdf.

B.5. Cargo Handling Equipment

This section describes the methodology used for determining BAU CHE emissions from port-related activity.

B.5.1. Modeling Approach

The baseline 2011 CHE sector activity was grown to 2020 and 2030 using the commodity growth rates shown in Table B-2. Emission factors were calculated from the baseline inventories, adjusted for expected changes in the future fleet, and then applied to the projected activity to determine the BAU inventories.

Gross emission factors[90] were calculated from the baseline 2011 CHE inventory using the following equation:

$$EF_{2011} = E_{2011}/(C_{2011} \times S)$$

Eq. B-6

Where:

EF = Emission factor for a specific pollutant and port (g/ton),

E = Total annual emissions for a specific pollutant and port (g), and

C = Total cargo throughput for a specific port (tons).

To calculate the gross emission factors, the total annual emissions came from the baseline CHE inventories (see Appendix A). The total cargo throughput came from USACE's Waterborne Commerce Statistics.[91] However, since fleets turn over to newer models in future years that meet stricter emission standards, the gross emission factors were scaled for use in 2020 and 2030 based on projected future fleet emission factors obtained from national scale runs of EPA's NONROAD model.[92] Emission factors for all CHE were determined by dividing the sum of national emissions for port-related cargo handling equipment (a list of applicable SCCs is given in Table B-6) by the sum of the national populations of the same equipment. These emission factors calculated from the model runs were not used directly because they are in terms of grams per vehicle per year, and the units required here were grams per ton of cargo moved. However, their values as used in this scaling analysis are presented in Table B-7. Please note that this technique was only used for NOx, $PM_{2.5}$, and HC inventories; the other pollutants were calculated separately.

---

[90] Used here, a "gross emission factor" is an estimate of emissions per unit goods moved. It is described as "gross" to distinguish from more refined factors, such as engine-, equipment-, operations-, or technology-specific emission factors determined from sources such as engine certification or emissions models.

[91] Available at: http://www.navigationdatacenter.us/db/wcsc/archive/xls/man11/.

[92] Available at: https://www.epa.gov/otaq/nonrdmdl.htm.

Table B-6. CHE Types by SCC from NONROAD

SCC	Equipment Type	Equipment Category
2270002015	Rollers	Construction and Mining Equipment
2270002027	Signal Boards/Light Plants	Construction and Mining Equipment
2270002036	Excavators	Construction and Mining Equipment
2270002045	Cranes	Construction and Mining Equipment
2270002051	Off-highway Trucks	Construction and Mining Equipment
2270002060	Rubber Tire Loaders	Construction and Mining Equipment
2270002066	Tractors/Loaders/Backhoes	Construction and Mining Equipment
2270002069	Crawler Tractor/Dozers	Construction and Mining Equipment
2270002072	Skid Steer Loaders	Construction and Mining Equipment
2270002075	Off-highway Tractors	Construction and Mining Equipment
2270003010	Aerial Lifts	Industrial Equipment
2270003020	Forklifts	Industrial Equipment
2270003030	Sweepers/Scrubbers	Industrial Equipment
2270003050	Other Material Handling Equipment	Industrial Equipment
2270003070	Terminal Tractors	Industrial Equipment
2270006005	Generator Sets	Commercial Equipment
2270006010	Pumps	Commercial Equipment
2270006015	Air Compressors	Commercial Equipment
2270006025	Welders	Commercial Equipment

Table B-7. Projected Emission Factors (Annual Grams per Equipment) from NONROAD

Calendar Year	NOx	HC	PM
2011	57,909	14,505	4,660
2020	23,268	6,816	1,817
2030	14,667	5,692	830

The future year gross emission factors were calculated using the following equations:

$$EF_{2020} = EF_{2011} \times \left[\frac{PopEF_{2020}}{PopEF_{2011}}\right] \qquad \text{Eq. B-7}$$

$$EF_{2030} = EF_{2011} \times \left[\frac{PopEF_{2030}}{PopEF_{2011}}\right] \qquad \text{Eq. B-8}$$

Where:

EF = Emission factor for a specific pollutant and port (g/ton) and

PopEF = Population based emission factor for a specific pollutant (Table B-7).

To calculate the BAU inventories, the future year emission factors were combined with grown CHE activity as shown in the following equations:

$$E_{2020} = EF_{2020} \times C_{2011} \times (1 + G)^{(2020-2011)} \qquad \text{Eq. B-9}$$

$$E_{2030} = EF_{2030} \times C_{2011} \times (1 + G)^{(2030-2011)}$$

Eq. B-10

Where:

E = Total annual emissions for a specific pollutant and port (g),

EF = Emission factor for a specific pollutant and port (g/ton),

C = Total cargo throughput for a specific port (tons), and

G = Commodity-based CAGR for the region in which the port is located (see Table B-2).

B.5.2. Additional Pollutants

The methodology described above provides inventories for NOx, $PM_{2.5}$, and HC only; therefore, VOC, BC, CO_2, benzene, acetaldehyde, and formaldehyde needed be calculated separately.

B.5.2.1. VOC

The future year proportional changes in VOC emissions were assumed to be equal to the proportional changes in HC emissions.

B.5.2.2. BC

BC emissions were estimated to be 0.77 times the $PM_{2.5}$ emissions, which is consistent with EPA's Report to Congress on Black Carbon.[93]

B.5.2.3. CO_2

Future year CO_2 emission factors were assumed to remain constant and equal to the baseline year. For the 2050 CO_2 inventory, the 2030 CAGR were applied to grow the cargo throughput values to 2050 levels.

B.5.2.4. Air Toxics

Emissions for the air toxics formaldehyde, benzene, and acetaldehyde were estimated as a fraction of VOC emissions based on speciation profiles calculated from running EPA's NMIM.[94] These speciation profiles are shown in Table B-8.

Table B-8. CHE Air Toxic Speciation Profiles from VOC Based on NMIM

Pollutant	2020	2030
Acetaldehyde	0.0155	0.0126
Benzene	0.0277	0.0286
Formaldehyde	0.0333	0.0268

---

[93] U.S. Environmental Protection Agency, *Report to Congress on Black Carbon*, EPA-450/R-12-001, March 2012, p. 87.

[94] Available at: https://www.epa.gov/otaq/nmim.htm.

B.6. Harbor Craft

This section describes the methodology used for determining BAU harbor craft emissions.

B.6.1. Modeling Approach

The projected 2020 and 2030 BAU emission inventories were developed as the product of emission factors and activity data. To facilitate this, the sector was split into two categories: goods-moving and non-goods moving. For vessels directly tied to goods movement, such as the various categories of tug, tow, and push, the activity growth was grown to 2020 and 2030 using the commodity growth rates shown in Table B-2.

For goods-moving harbor craft, gross emission factors[95] were calculated from the baseline 2011 harbor craft inventory using the following equation:

$$EF_{2011} = E_{2011}/C_{2011}$$ Eq. B-11

Where:

EF = Goods-moving emission factor for a specific pollutant and port (g/ton),

E = Goods-moving annual emissions for a specific pollutant and port (g), and

C = Total cargo throughput for a specific port (tons).

To calculate the gross emission factors for goods-moving harbor craft, the goods-moving annual emissions came from the baseline harbor craft inventories (see Appendix A). The total cargo throughput came from USACE's Waterborne Commerce Statistics.[96] Combining these results in gross emission factors that are valid for the 2011 goods-moving harbor craft fleet at each port.

For non-goods moving harbor craft, gross fuel-based emission factors were calculated from the baseline inventory using the following equation:

$$EF_{2011} = E_{2011}/FC_{2011}$$ Eq. B-12

Where:

EF = Non-goods moving emission factor for a specific pollutant and port (g/gallon),

E = Non-goods moving annual emissions for a specific pollutant and port (g), and

FC = Non-goods moving annual fuel consumption (gallons).

---

[95] Used here, a "gross emission factor" is an estimate of emissions per unit goods moved. It is described as "gross" to distinguish from more refined factors, such as engine-, equipment-, operations-, or technology-specific emission factors determined from sources such as engine certification or emissions models.

[96] Available at: http://www.navigationdatacenter.us/db/wcsc/archive/xls/man11/.

The non-goods moving portion of the total annual emissions came from the baseline inventory for vessel types such as ferries, support, fishing, and government. The fuel consumption was estimated from the non-goods moving baseline CO_2 inventories: E_{CO_2} [g] / (26.34% [fuel carbon content] * 3207 [g/gal] * 3.664 [CO_2 to C ratio]). Combining these results in gross emission factors that are valid for the 2011 non-goods moving harbor craft fleet at each port.

However, since fleets turn over to newer models in future years that meet stricter emission standards, both sets of emission factors needed to be adjusted. Therefore, the gross emission factors for both goods-moving and non-goods moving were scaled for use in 2020 and 2030 based on projected emissions per vessel as calculated from EPA's 2008 Locomotive and Marine Emission Standards Rulemaking.[97] Total emissions for the control case for 2020 and 2030 were divided by the total number of C1 and C2 engines in those years to determine average emissions of each pollutant per marine engine.

The emission factors calculated from the rulemaking were not used directly because they varied by age, engine type (main or auxiliary), engine power, and engine displacement; this level of disaggregation was impractical to repeat for this activity. As such, the calculated values as used in this scaling analysis are presented in Table B-9. Please note that only NOx, $PM_{2.5}$, and VOC were available for this technique; the other pollutants were calculated separately.

Table B-9. Projected Emission Factors (Annual Grams per Marine Engine) from 2008 Locomotive and Marine RIA

Calendar Year	NOx	VOC	$PM_{2.5}$
2011	5,654,548	123,910	198,748
2020	3,592,204	79,765	111,930
2030	2,022,078	43,223	64,705

Future year gross emission factors were calculated using the following equations:

$$EF_{2020} = EF_{2011} \times \left[\frac{PopEF_{2020}}{PopEF_{2011}}\right]$$ Eq. B-13

$$EF_{2030} = EF_{2011} \times \left[\frac{FuelEF_{2030}}{FuelEF_{2011}}\right]$$ Eq. B-14

Where:

EF = Emission factor for a specific pollutant and port (g/ton) and

PopEF = Population based emission factor for a specific pollutant (Table B-9).

---

[97] U.S. Environmental Protection Agency, *Control of Emissions of Air Pollution from Locomotive Engines and Marine Compression Ignition Engines Less than 30 Liters per Cylinder*, EPA420-R-08-001, March 2008.

To calculate the goods-moving BAU inventories, the future year emission factors were combined with grown cargo tonnage moved at each port as shown in the following equations:

$$E_{2020} = EF_{2020} \times C_{2011} \times (1 + G)^{(2020-2011)}$$ **Eq. B-15**

$$E_{2030} = EF_{2030} \times C_{2011} \times (1 + G)^{(2030-2011)}$$ **Eq. B-16**

Where:

E = Goods-moving emissions for a specific pollutant and port (g),

EF = Goods-moving emission factor for a specific pollutant and port (g/ton),

C = Total cargo throughput for a specific port (tons), and

G = Commodity-based CAGR for the region in which the port is located (see Table B-2).

For the non-goods moving BAU inventories, the activity was assumed to be inelastic to changes in cargo movement and therefore assumed to have no growth. In other words, vessels such as ferries, support (offshore & research), and fishing were assumed to operate at 2011 activity levels for all future years. This assumption was used in lieu of better data as no nationwide projected values for harbor craft are readily available. The 2008 study by the Research Triangle Institute, the basis of the growth rates shown in Table B-2, discusses global historical and projected fuel consumption for non-goods moving vessels but does not provide domestic values. The California Air Resources Board (ARB) OFFROAD model predicts no change in the statewide fleet and fuel consumption for crew and supply vessels for all years through 2025 and no change in the number of commercial fishing vessels from 2009 through 2030, while showing a substantial increase in the number of "commercial" boats. However, it also predicts the number of tugs to remain constant over this period. This approach is also similar to that used in the Energy Information Administration's (EIA's) National Energy Modeling System, which determines fuel demand for goods movement activities in the Freight Transport Module in two main categories: freight and recreational. However, those categories are not good matches to those used here and NEI projections are not likely to be consistent with the factors used here for vessels involved in goods movements.

Therefore, to calculate the non-goods moving BAU inventories, the future year emission factors were combined with the 2011 non-goods moving fuel consumption at each port as shown in the following equations:

$$E_{2020} = EF_{2020} \times FC_{2011}$$ **Eq. B-17**

$$E_{2030} = EF_{2030} \times FC_{2011}$$ **Eq. B-18**

Where:

E = Goods-moving emissions for a specific pollutant and port (g),

EF = Goods-moving emission factor for a specific pollutant and port (g/gallon), and

FC = Non-goods moving 2011 fuel consumption (gallons).

B.6.2. Additional Pollutants

The methodology described above provides inventories for NOx, PM$_{2.5}$, and VOC only; therefore, BC, CO$_2$, benzene, acetaldehyde, and formaldehyde needed be calculated separately.

B.6.2.1. BC

BC emissions were estimated to be 0.77 times the PM$_{2.5}$ emissions, which is consistent with EPA's Report to Congress on Black Carbon.[98]

B.6.2.2. CO$_2$

Future year CO$_2$ emission factors were assumed to remain constant and equal to the baseline year. For the goods-moving 2050 CO$_2$ inventory, the 2030 CAGR were applied to grow the cargo throughput values to 2050 levels. For the non-goods moving 2050 inventory, the 2011 non-goods moving activity values were used.

B.6.2.3. Air Toxics

Emissions for the air toxics formaldehyde, benzene, and acetaldehyde were estimated as a fraction of VOC emissions based on diesel speciation profiles calculated from running MOVES2010b. MOVES was used for the harbor craft sector due to a lack of other sources more directly applicable to harbor craft: NONROAD does not include commercial marine and SPECIATE and the NEI do not include projections. The MOVES speciation profiles are shown in Table B-10. These fractions vary depending on whether or not VOC is controlled (model year 2007 and later). For marine engines, this relates primarily to engines mandated to use Tier 4 emission controls. Weighted speciation factors were calculated for 2020 and 2030 based on Tier 4 versus pre-Tier 4 engine distributions from the 2008 Locomotive and Marine RIA.[99] This is not consistent with the MSAT Rule or the Emissions Modeling Platform, which use constant speciation factors for future years.

Table B-10. Diesel Truck Air Toxic Speciation Profiles Based on MOVES2010b Applied to Harbor Craft

Pollutant	Toxic/VOC without control	Toxic/VOC with Control
Acetaldehyde	0.035559	0.06934
Benzene	0.007835	0.01291
Formaldehyde	0.078225	0.21744

---

[98] U.S. Environmental Protection Agency, *Report to Congress on Black Carbon*, EPA-450/R-12-001, March 2012, p. 87.

[99] U.S. Environmental Protection Agency, *Control of Emissions of Air Pollution from Locomotive Engines and Marine Compression Ignition Engines Less than 30 Liters Per Cylinder*, EPA420-R-08-001, March 2008.

B.7. Ocean Going Vessels

This section describes the methodology used for determining BAU OGV emissions.

B.7.1. Modeling Approach

The business as usual inventories for OGV are based primarily upon the methodology used for the Category 3 Marine Engine Rulemaking[100] (C3 RIA). Using the C3 RIA modeling approach, the OGV emission inventories were calculated using energy-based emission factors combined with activity profiles for vessels calling at each port.

Essentially, the baseline 2011 number of calls at each port included in this assessment was grown to 2020 and 2030 using the annual average growth rate as listed in Table B-1 for the appropriate region. This action alone would proportionally increase each of the inventories. However, adjustment factors were also applied to the calculations for NOx, PM, BC, SO$_2$ and CO$_2$ to account for fleet turnover and lower sulfur fuels.

The IMO adopted NOx limits in Annex VI to the International Convention for Prevention of Pollution from Ships in 1997. These NOx limits apply for all marine engines over 130 kW for engines built on or after January 1, 2000, including those that underwent a major rebuild after January 1, 2000. For the C3 RIA, EPA determined the effect of the IMO standard to be a reduction in NOx emission rate of 11% below that the standard for engines built before 2000. Therefore, for engines built between 2000 and 2010 (Tier I), a NOx emission adjustment of 0.89 was applied to the calculation of NOx emissions for both propulsion and auxiliary engines. IMO Tier II NOx emission standards came into effect in 2011 and represent a 2.5 g/kWh reduction over Tier I engines. Tier III came into effect for engines built in 2016, which represents an 80% reduction from Tier I. Thus Tier III emission factors are 20% of Tier I emission factors. All emission factors are consistent with the C3 RIA.

In addition to the MARPOL Annex VI emission limits that apply to all ships engaged in international transportation, US vessels must also comply with EPA's Clean Air Act requirements for engines and fuels. The NOx emission limits for Category 3 engines are equivalent to the MARPOL Annex VI NOx limits. EPA's sulfur limit for distillate locomotive or marine (LM) diesel fuel sold in the United States is more stringent (15 ppm sulfur) than the ECA fuel sulfur limit (1000 ppm sulfur starting 2015); the sulfur limit for ECA fuel for use on Category 3 marine vessels is equivalent to the MARPOL Annex VI SOx limits. EPA also has emission standards for C3 engines,[101] which are generally the same or more stringent but almost all C3 engines used in international shipping fall under IMO regulations.

---

[100] U.S. Environmental Protection Agency, *Regulatory Impact Analysis: Control of Emissions of Air Pollution from Category 3 Marine Diesel Engines,* EPA Report EPA-420-R-09-019, December 2009. Available at: http://www.epa.gov/otaq/regs/nonroad/marine/ci/420r09019.pdf.

[101] U.S. Environmental Protection Agency, *Control of Emissions from New Marine Compression-Ignition Engines at or Above 30 Liters per Cylinder,* Federal Register, Vol 75, No 83, April 30, 2010.

In addition, as part of the new IMO standards, marine diesel engines built between 1990 and 1999 that are 90 liters per cylinder or more need to be retrofitted by 2020 to meet Tier 1 emission standards upon engine rebuild if a retrofit kit is available to the ships. Consistent with the C3 RIA, it was assumed that 80% of all ships > 90L / cylinder will have retrofit kits available.

To calculate NOx reductions due to fleet turnover, NOx adjustment factors were calculated for 2020 and 2030 based upon all ports examined in this analysis combined. This simplification allowed the same factors to be applied to all 19 ports. To accomplish this, installed power age profiles by engine type for propulsion engines and by vessel type for auxiliary engines were developed using the 2011 Entrances and Clearances data and Lloyd's vessel characterization data. For propulsion engines, installed power by engine type was calculated for each model year based upon the sum of the total propulsion power of Category 3 vessels over the Entrances and Clearances data. In addition, to calculate the effect of retrofitting Tier 0 engines of more than 90 liters per cylinder, installed power was also calculated for slow speed diesel (SSD) and medium speed diesel (MSD) engines that were over 90 liters per cylinder. Ages were determined by subtracting the build year from 2011. This 2011 age profile was then used in both 2020 and 2030 adjusting model years to fit the age profile. For example, a five-year-old engine in 2011 is a 2006 model year, but in 2020 is a 2015 model year and in 2030 a 2025 model year. This same methodology was used in the C3 RIA.

For auxiliary engines, auxiliary power was calculated from the propulsion power using the auxiliary power to propulsion power ratios by ship type found in Table A-20. This is a slight variation from the C3 RIA, which instead used the propulsion installed power to calculate auxiliary engine NOx adjustment factors. Auxiliary engines were only segregated into passenger ships and other ships because in 2011 different residual oil (RO) to marine gas oil (MGO) ratios were used. The installed power by age and engine type (or vessel type for auxiliary engines) is shown in Table B-11.

Table B-11. 2011 Installed Power by Engine Type (kW)

Age (yrs)	Propulsion Engines				Auxiliary Engines	
	MSD	SSD	GT	ST	Passenger	Other
58	-	883	-	-	-	238
56	4,415	155,328	-	-	43,181	1,188
54	60,044	-	-	-	-	16,152
51	5,880	-	-	-	-	1,123
47	-	-	-	28,318	-	6,287
46	1,104	-	-	-	-	211
45	5,663	-	-	-	-	1,082
44	23,055	-	-	11,400	-	7,562
43	23,536	-	-	-	6,543	-
41	3,884	-	-	-	-	742
40	142,385	156,512	-	235,360	-	118,761
39	683,299	-	-	-	-	171,013
38	40,578	52,950	-	94,144	9,812	33,484
37	13,245	96,350	-	23,536	-	26,111
36	113,264	237,351	-	-	-	69,933

Age (yrs)	Propulsion Engines				Auxiliary Engines	
	MSD	SSD	GT	ST	Passenger	Other
35	61,451	41,192	-	-	-	25,096
34	683,005	818,491	-	-	-	330,714
33	419,242	962,313	-	63,254	-	326,991
32	1,416,412	1,307,594	-	-	-	691,957
31	1,289,949	1,589,319	-	-	-	674,436
30	168,980	3,951,662	-	-	9,681	895,328
29	1,583,342	2,503,409	-	-	-	918,034
28	102,456	6,291,171	-	-	-	1,571,537
27	36,852	17,197,530	-	-	-	3,927,271
26	235,329	7,271,460	-	-	-	1,694,493
25	53,936	8,010,446	-	-	-	1,806,526
24	134,791	6,790,457	-	-	-	1,590,847
23	170,487	11,922,756	-	-	11,743	2,699,496
22	2,316,184	14,011,939	-	-	636,028	3,150,791
21	6,409,732	4,072,808	-	-	1,616,225	1,081,023
20	1,341,768	14,379,033	-	-	299,533	3,252,828
19	4,850,116	12,301,414	-	-	1,262,876	2,858,540
18	537,682	22,408,866	-	-	76,861	5,494,131
17	1,824,020	28,760,465	-	-	413,430	6,973,053
16	12,471,612	44,043,180	-	-	3,402,311	9,750,108
15	18,714,120	50,532,802	-	-	5,067,754	11,270,238
14	10,139,100	55,342,547	-	-	2,411,583	12,629,186
13	25,079,400	49,681,652	-	-	6,361,730	11,478,444
12	23,817,894	32,483,814	-	-	5,965,327	7,798,999
11	23,603,136	60,132,802	4,701,972	-	6,975,766	13,978,348
10	27,380,891	65,091,176	13,058,032	-	10,594,636	14,892,250
9	13,195,720	87,355,544	-	105,920	3,083,740	19,632,342
8	16,670,650	75,944,446	10,315,048	211,824	6,921,840	17,047,496
7	31,990,234	85,901,288	1,265,000	116,208	8,168,129	19,686,618
6	22,460,220	97,966,162	-	-	5,752,821	21,910,074
5	14,017,483	111,010,740	-	116,208	2,196,800	25,779,985
4	18,001,132	129,071,986	-	107,104	3,482,793	29,685,530
3	18,759,733	103,070,751	-	57,330	3,682,955	24,017,466
2	3,278,815	86,978,426	-	55,200	349,279	19,553,144
1	13,620,786	93,117,757	-	52,992	3,199,891	21,069,329
0	2,370,454	24,647,140	-	-	557,446	5,617,455

Emission factors for propulsion engines by engine type, fuel type, and emission tier are shown in Table B-12. This table combines the base emission factors as discussed in section A.6.6 and these adjustments discussed above.

Table B-12. NOx Emission Factors by Engine Type

Engine Type	Tier	Engine Model Years	Emission Factor (g/kWh)	
			RO	MDO/MGO
MSD	0	Pre-2000	14.0	13.2
	1	2000 -2010	12.5	11.7
	2	2011 -2015	10.0	9.2
	3	2016+	--	2.3
SSD	0	Pre-2000	18.1	17.0
	1	2000 -2010	16.1	15.1
	2	2011 -2015	13.6	12.6
	3	2016+	--	3.0
GT	0	All	6.1	5.7
ST	0	All	2.1	2.0
Auxiliary	0	Pre-2000	14.7	13.9
	1	2000 -2010	13.1	12.4
	2	2011 -2015	10.6	9.9
	3	2016+	--	2.5

An ARB survey published in 2005[102] found that almost all ships used RO in their main propulsion engines, and that only 29% of all ships (except passenger ships) used distillate (MGO/MDO) in their auxiliary engines, with the remaining 71% using RO. Only 8% of passenger ships used distillate in their auxiliary engines, while the other 92% used RO. Even though these two fuels are not blended in any given vessel, the emission factor used in the analysis represents an average of the two fuels, weighted by the relative market share of each. For all other ships, 29% used distillate and 71% used RO. Table B-13 shows the NOx emission factors by Tier for the two ship types.

Table B-13. Auxiliary Engine NOx Emission Factors by Ship Type

Calendar Year	Tier	Engine MY	Emission Factor (g/kWh)	
			Passenger	Other
2011	0	Pre-2000	14.6	14.5
	1	2000 -2010	13.0	12.9
	2	2011	10.5	10.4
2020 and later	0	Pre-2000	13.9	13.9
	1	2000 -2010	12.4	12.4
	2	2011 -2015	9.9	9.9
	3	2016+	2.5	2.5

---

[102] California Air Resources Board, *2005 Oceangoing Ship Survey, Summary of Results,* September 2005

Using the above information, average NOx emission factors were calculated for the years 2020 and 2030. Engine model years were assigned to the age profile based upon calendar year. For 2020, age 0 was model year 2020 and for 2030, age 0 was model year 2030. Next, emission tiers and NOx emission factors from Table B-12 were assigned to the various model years. Then, the effect of the retrofit requirement for engines built between 1990 and 1999 were taken into account. Average NOx emission factors were determined for 2011, 2020, and 2030 by taking the sum of the installed power times the emission factor for each model year and dividing the sum-product by the sum of the installed power and the results are shown in Table B-14. A similar process was used for auxiliary engines, and the resulting average NOx emission factors are given in Table B-15. This is different from the C3 RIA, where age distributions were broken out by Great Lakes/Deep Sea Ports instead of ship type. NOx adjustment factors were then calculated by dividing the 2020 or 2030 average NOx emission factor by the 2011 average NOx emission factor for a given engine or vessel type. These are given in Table B-16. These adjustment factors were then applied to the calculation of NOx emissions by engine type and port.

Table B-14. Average Propulsion Engine NOx Emission Factor (g/kWh) by Engine Type

Year	MSD	SSD	GT	ST
2011	13.0	16.6	6.1	2.1
2020	9.4	10.6	5.7	2.0
2030	3.7	5.0	5.7	2.0

Table B-15. Average Auxilliary Engine NOx Emission Factor (g/kWh) by Ship Type

Year	Passenger	Other
2011	13.6	13.3
2020	10.3	8.6
2030	3.7	4.1

Table B-16. NOx Adjustment Factors

Year	Propulsion Engines				Auxiliary Engines	
	MSD	SSD	GT	ST	Passenger	Other
2020	0.7233	0.6389	0.9344	0.9524	0.7582	0.6505
2030	0.2856	0.2988	0.9344	0.9524	0.2736	0.3064

B.7.2. Additional Pollutants

Fuel changes from 2.7% sulfur RO to 0.1% sulfur MDO starting in 2015 affects PM_{10}, $PM_{2.5}$, BC, SO_2 and CO_2 emissions. Emission factors for the two fuels used in propulsion engines are shown in Table B-17 and come from Entec.[103] Please note that in the C3 RIA, different RO sulfur levels were used for West

---

[103] Entec UK Limited, *Quantification of Emissions from Ships Associated with Ship Movements between Ports in the European Community*, prepared for the European Commission, July 2002.

Coast ports. Since California ports were not included in this analysis and the 2011 Starcrest Inventory for Port of Seattle used 2.7% sulfur for RO, we have used 2.7% sulfur RO for all ports.

Table B-17. Average Propulsion Engine Emission Factors (g/kWh) by Engine Type

Engine Type	Fuel Type	Sulfur	Emission Factor (g/kWh)					
			PM_{10}	$PM_{2.5}$	BC	SO_2	CO_2	BSFC
SSD	RO	2.70%	1.42	1.31	0.039	10.29	621	195
	MDO	0.10%	0.19	0.17	0.010	0.36	589	185
MSD	RO	2.70%	1.43	1.32	0.040	11.24	678	213
	MDO	0.10%	0.19	0.17	0.010	0.40	646	203
GT	RO	2.70%	1.47	1.35	0.040	16.10	971	305
	MDO	0.10%	0.17	0.15	0.009	0.57	923	290
ST	RO	2.70%	1.47	1.35	0.040	16.10	971	305
	MDO	0.10%	0.17	0.15	0.009	0.57	923	290

The fuel changes also affect the same emissions for auxiliary engines. Emission factors for the two fuels used in auxiliary engines are shown in Table B-18 (also from Entec). BSFC from Entec was used based upon the ratio of RO versus MDO listed in the table.

Table B-18. Average Auxilliary Engine Emission Factors (g/kWh) by Ship Type

Ship Type	Fuel Type	Sulfur	Emission Factor (g/kWh)					
			PM_{10}	$PM_{2.5}$	BC	SO_2	CO_2	BSFC
Passenger	92% RO/8% Distillate	2.56%	1.36	1.25	0.038	11.36	718	226
	100% MGO	0.10%	0.18	0.17	0.010	0.42	691	217
Other	71% RO/29% Distillate	2.21%	1.16	1.07	0.032	9.74	707	224
	100% MGO	0.10%	0.18	0.17	0.010	0.42	691	217

Adjustment factors taking into account the change in emission factors for both propulsion and auxiliary engines due to lower sulfur fuel are shown in Table B-19 and were applied to emissions calculations for both 2020 and 2030. It was assumed that by 2020 and 2030, all vessels will be using distillate fuel at these ports.

Table B-19. Fuel Adjustment Factors for 2020 and 2030

Engine	Type	PM_{10}	$PM_{2.5}$	BC	SO_2	CO_2
Propulsion	MSD	0.1329	0.1329	0.2659	0.0351	0.9487
	SSD	0.1295	0.1295	0.2591	0.0353	0.9531
	GT	0.1134	0.1134	0.2268	0.0352	0.9508
	ST	0.1134	0.1134	0.2268	0.0352	0.9508
Auxiliary	Passenger	0.1340	0.1340	0.2680	0.0373	0.9617
	Other	0.1569	0.1569	0.3138	0.0436	0.9772

Finally, it was assumed that there would be an increase in the use of shore power at a limited number of ports in this assessment. It was assumed that those ships that use shore power only emit auxiliary engine emissions during the time the shore power cables are being connected to and disconnected from the ship. This was estimated to take two hours per call. Assuming an average of

10 hours hoteling per call for passenger ships results in an 80% reduction in all emissions except CO_2 that use shore power. CO_2 emissions from the power plant that generates electricity must also be considered. Based upon current projections of U.S. average generation mix[104] using the Argonne National Laboratories GREET2014 model,[105] electricity generation produces 517 g/kWh of CO_2 at the plug compared for 2020, 477 g/kWh in 2030 and 461 g/kWh in 2050[106] to the 691 g/kWh generated by the auxiliary engines. Taking into account the time the cables are connected, this results in a 20, 25 and 27% reduction in CO_2 emissions for 2020, 2030 and 2050, respectively, for ships that use shore power.[107]

B.8. BAU Summary Results

The following figures show the 2011 baseline inventories combined with the 2020 and 2030 business as usual projections for each pollutant, aggregated by sector. As noted earlier, SO_2 was only calculated for OGV and acetaldehyde, benzene, and formaldehyde were only calculated for the non-OGV sectors.

Figure B-1. Total NOx Emissions Aggregated by Sector, Tons/Year

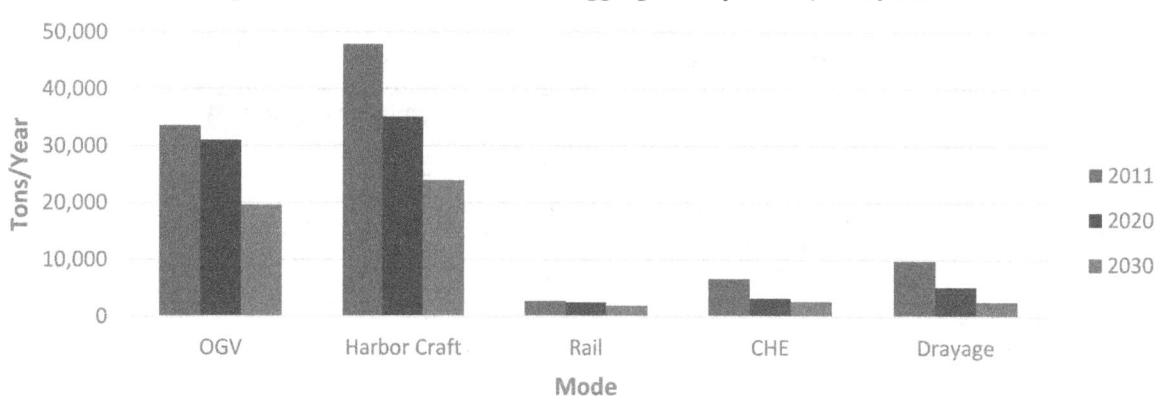

---

[104] The CO_2 emission rates calculated using GREET2014 assume an average U.S. generation mix. At some ports, the generation mix is significantly different and would thus have a different emission factor. For example, in the Northwest, much of the electricity comes from hydropower therefore utilities emit less CO_2 overall.

[105] Argonne National Laboratories, *GREET Model 2014.* Available at: https://greet.es.anl.gov/.

[106] GREET2014 only extrapolates to 2040, so the 2040 CO_2 emission rate for power plants was also used for 2050.

[107] Note that cargo loading and unloading occur while the connection is being made and removed, so the total hoteling time estimate is expected to be unchanged by shore power, although on a first call at a new terminal commissioning is required, which takes much longer. This calculation assumes 2 hours per call for connection and disconnection, but does not include any commissioning time.

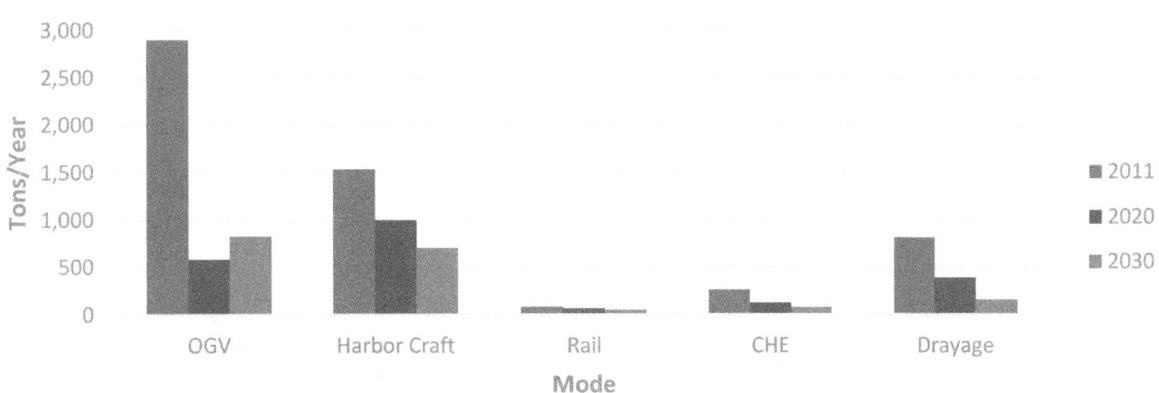

Figure B-2. Total PM$_{2.5}$ Emissions Aggregated by Sector, Tons/Year

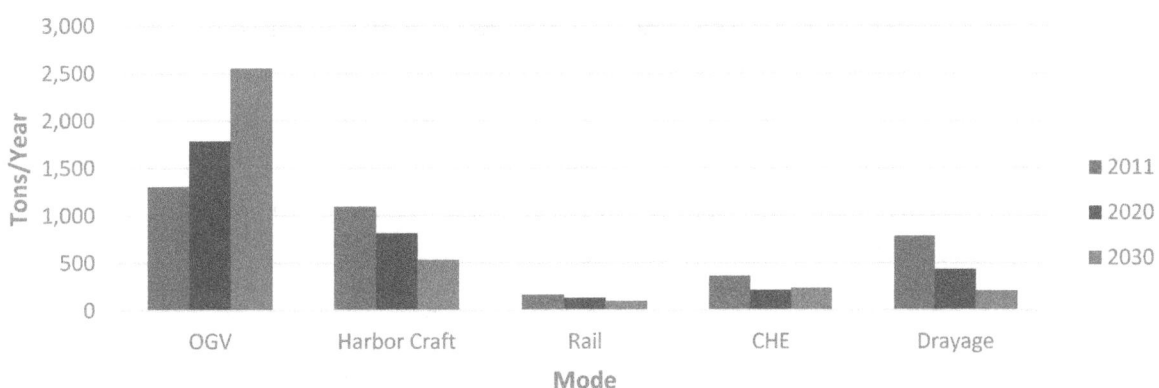

Figure B-3. Total VOC Emissions Aggregated by Sector, Tons/Year

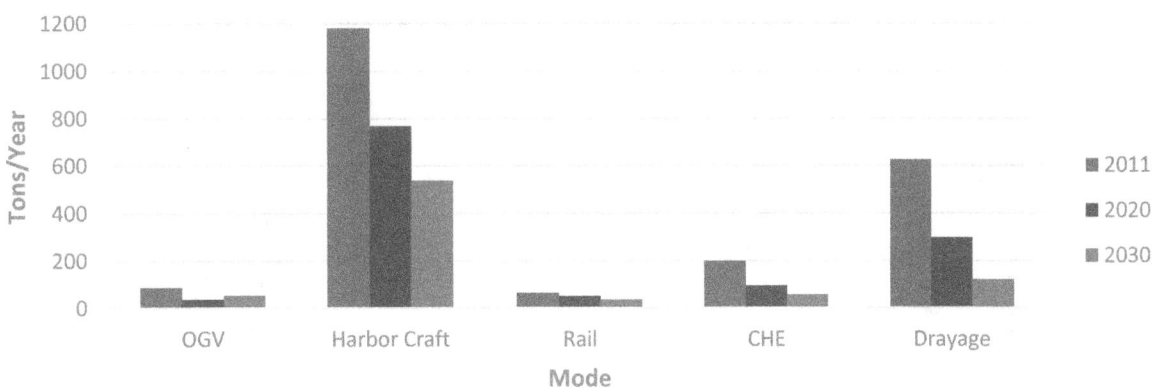

Figure B-4. Total BC Emissions Aggregated by Sector, Tons/Year

Figure B-5. Total CO₂ Emissions Aggregated by Sector, Tons/Year

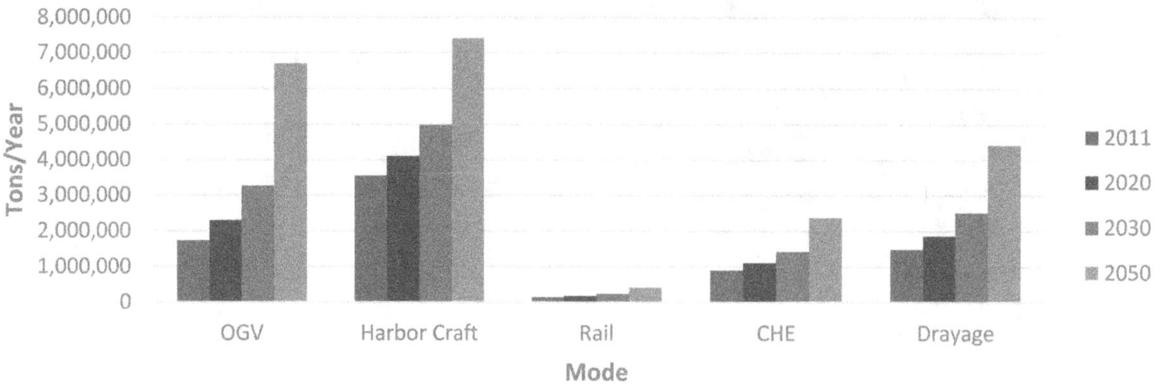

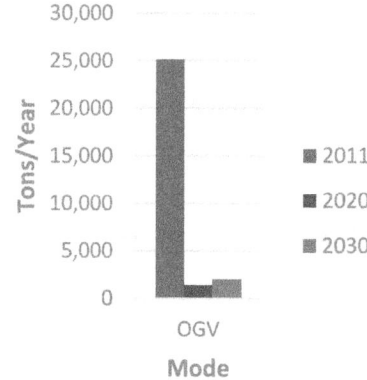

Figure B-6. Total SO₂ OGV Emissions Aggregated, Tons/Year

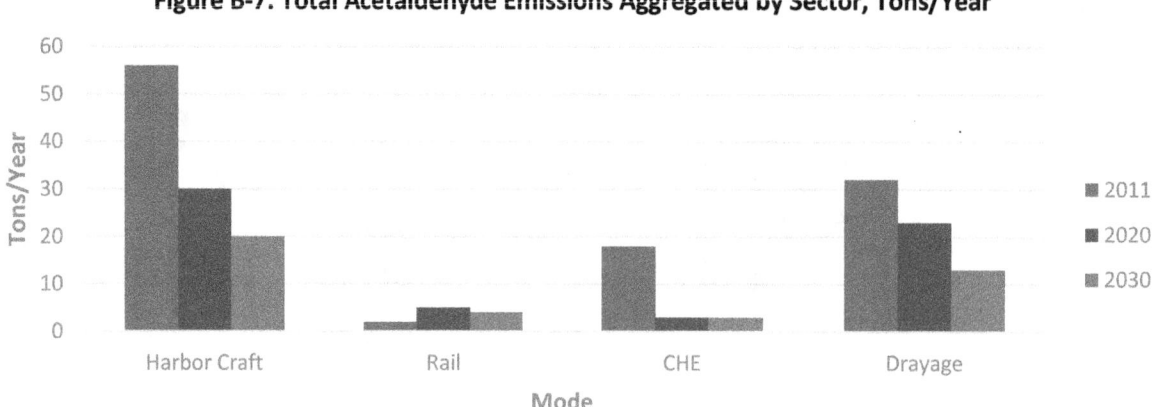

Figure B-7. Total Acetaldehyde Emissions Aggregated by Sector, Tons/Year

Figure B-8. Total Benzene Emissions Aggregated by Sector, Tons/Year

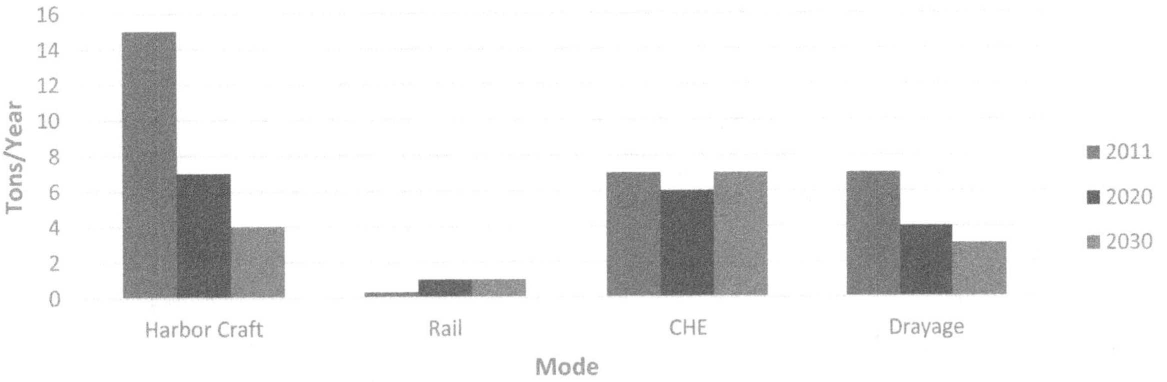

Figure B-9. Total Formaldehyde Emissions Aggregated by Sector, Tons/Year

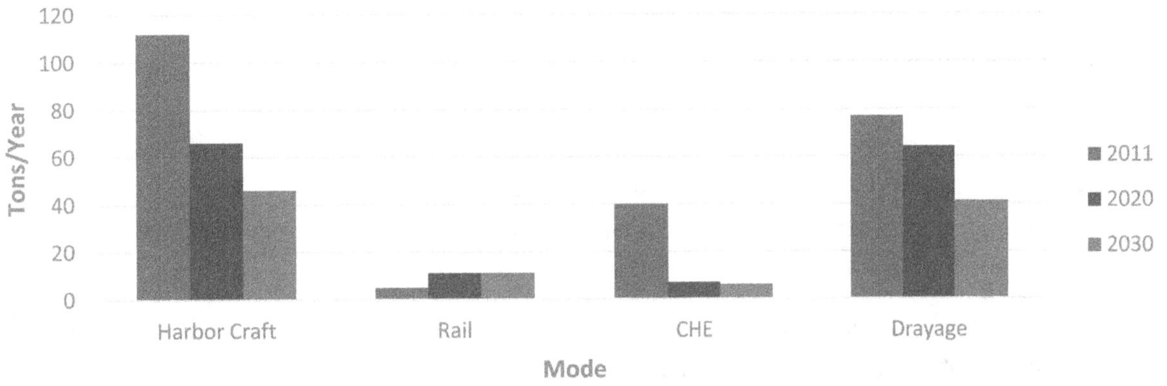

Appendix C.   Analysis of Emission Reduction Scenarios

C.1. Overview

This appendix describes the methodology and other assumptions that support the scenario analysis results presented in Section 6 of the final report.

C.1.1. Intent of Scenarios

The strategy scenarios were developed based on the screening-level assessment described in Section 5 where a range of potential technological and operational strategies were evaluated, in addition to EPA's existing expertise on port-related strategies and consultation with port stakeholder experts. Strategy scenarios were developed for each mobile source sector for the years 2020 and 2030 for all pollutants and for only CO_2 in 2050. Although the specific strategies differ between sectors, the purpose of all strategy scenarios are as follows:

- **Scenario A** was intended to reflect an increase in the introduction of newer technologies in port vehicles and equipment beyond what would occur through normal fleet turnover. Operational strategies in Scenario A reflect a reasonable increase in expected efficiency improvements for drayage truck, rail, and OGV sectors. For the OGV sector, moderate levels of fuel switching and other emission control strategies are also analyzed. All of the strategies included in Scenario A may be supported by a moderate increase in public and private funding.

- **Scenario B** reflected a more aggressive suite of strategies as compared to Scenario A. Scenario B would necessitate a major public and private investment to accelerate introduction of zero emissions vehicles, equipment, and vessels, in addition to different fuels and other technologies. Operational strategies in Scenario B assume further operational efficiency improvements beyond Scenario A.

In selecting strategies, EPA qualitatively considered several factors, such as strategy costs, implementation barriers, feasibility, and potential for market penetration by analysis year. However, an in-depth cost-benefit analysis was not conducted.

The remainder of this appendix provides more details regarding the methodology and assumptions used to estimate the strategy scenario reductions in Section 6.

C.2. Drayage Trucks

C.2.1. Scenarios

As discussed in Section 5, the scenarios were based on the future year distributions of drayage trucks consistent with the national default fleet turnover rates in EPA's MOVES2010b model, which is presented in Table C-1.

Table C-1. BAU Distribution of Trucks by Model Year

Model Year	2011	2020	2030	2050
pre-1991	20%	5%	0%	0%
1991-93	9%	6%	0%	0%
1994-97	21%	13%	0%	0%
1998-2003	24%	16%	7%	0%
2004-06	12%	9%	5%	0%
2007-09	10%	8%	5%	0%
2010+	4%	44%	84%	100%
Total	100%	100%	100%	100%

Table C-2 shows the strategy scenarios that were analyzed in this assessment. Note that model year references in this table indicate the emission standards that began with that model year (e.g., a 2007 truck means a truck that meets the EPA standards effective for model years 2007–2009).

Table C-2. Drayage Truck Strategy Scenarios

Strategy	2020/A	2020/B	2030/A	2030/B	2050/A	2050/B
Technological	Replace all pre-1994 trucks with 50% post-1998, 30% 2007, 20% 2010 or newer trucks	Replace all pre-1998 trucks with 50% 2007, 40% 2010, and 10% PHEV	Replace 100% of pre-2004 trucks with 2010 trucks. Replace 20% of 2004-09 trucks with PHEV	Replace 100% of pre-2007 trucks with 50% 2010 and 50% PHEV. Replace 10% of post-2010 with PHEV	Replace 25% of post-2010 trucks with PHEV	Replace 50% of post-2010 trucks with PHEV
Operational	Reduce gate queues by 25%	Reduce gate queues by 50%	Reduce gate queues by 25%	Reduce gate queues by 50%	Reduce gate queues by 25%	Reduce gate queues by 50%

Note that for the Operational Strategies, the percent of truck operating time spent in gate queues did not vary for the BAU inventories in 2020, 2030, and 2050. However, the number of drayage trucks increased with each future year, so the BAU emissions associated with gate queues also increased with each future year.

C.2.2. Technological Strategy Scenarios: Methodology and Assumptions

C.2.2.1. Relative Reduction Factors

The DrayFLEET model was used to determine a fleet-average relative reduction factor (RRF) for each scenario strategy. This is different from how the BAU inventories for drayage trucks were developed, where a default truck fleet age distribution from MOVES2010b was used in the DrayFLEET model. For the strategy scenario analysis, the BAU age distribution was replaced with alternative distributions consistent with each scenario. Since plug-in hybrid electric vehicles are not included in MOVES, these trucks were accounted for outside the model. For each scenario, a fleet-average RRF was calculated as the scenario emissions divided by the BAU emissions. To simplify calculations, a generic, national average fleet emissions RRF representing an average port (a "typical port") was used, rather than generating a separate model for each port.

Emission factors for conventional diesel drayage trucks for most pollutants were drawn from the emission standards reported in EPA's Heavy-Duty Highway Compression-Ignition Engines and Urban Buses Exhaust Emission Standards.[108] VOC emission rates were determined from the hydrocarbon (THC) standards, using EPA THC to VOC conversion factors.[109] Emission rates for the select air toxics acetaldehyde, benzene, and formaldehyde were calculated by applying speciation factors for diesel engines to VOC emissions. The analysis relied on EPA speciation profiles for on-road engines from the MOVES model.[110] Well-to-wheel CO_2 emission factors came from GREET 2015[111] using the 2015 HDV emission factor for CIDI Combination Short-Haul Conventional Diesel. For these calculations, no change in fuel economy or CO_2 emission rates were assumed between model year standards.[112] For PHEVs, emission factors were developed for NOx, $PM_{2.5}$, and CO_2 based on percent reduction values from SCAG's Comprehensive Regional Goods Movement Plan.[113] RRFs for VOCs were based on emission benefit percent reductions from the Gateway Cities Air Quality Action Plan.[114] RRFs for black carbon (BC) are assumed equal to that for $PM_{2.5}$ consistent with the BAU emission inventory methodology.

Because the DrayFLEET model cannot readily model each of these scenarios, the fleet-average emission factor was calculated from the emission standards described above weighted by truck population distributions specific to each scenario. To develop the truck population distributions for each scenario, the default truck fleet age distribution from the BAU methodology was used (originally taken from MOVES2010b). No PHEV trucks were assumed in the BAU fleet mixes. The population distributions were adjusted based on the scenarios. Table C-3 shows the model distributions that resulted from applying these scenarios to a hypothetical population of 1,000 trucks.

---

[108] U.S. Environmental Protection Agency, *Emission Standards Reference Guide*. Available at: http://www.epa.gov/otaq/standards/heavy-duty/hdci-exhaust.htm.

[109] U.S. Environmental Protection Agency, *Conversion Factors for Hydrocarbon Emission Components (July 2010)*, Report EPA-420-R-10-015 NR-002d.

[110] U.S. Environmental Protection Agency, *MOVES2010b: Additional Toxics Added to MOVES*. EPA-420-B-12-029a, May 2012, Sec 3.1.1. Available at: http://www.epa.gov/otaq/models/moves/documents/420b12029a.pdf.

[111] GREET2015 was released October 2, 2015. All calculations were updated to GREET2015 results. For information on the model, see https://greet.es.anl.gov/.

[112] Note that BAU inventories are based on DrayFLEET results. That model relies on a previous version of MOVES, which does not include EPA's heavy-duty engine and vehicle GHG regulations.

[113] ICF International, *Comprehensive Regional Goods Movement Plan, Task 10.2: Evaluation of Environmental Mitigation Strategies*, prepared for the Southern California Association of Governments, 2012.

[114] ICF International, *Gateway Cities Air Quality Action Plan, Task 7: New Measures Analysis*, prepare for the Los Angeles County Metropolitan Transportation Authority and the Gateway Cities Council of Governments, 2013.

Table C-3. Example of Drayage Truck Model Year Distribution by Scenario

Model Year	2020 BAU	2020/A	2020/B	2030 BAU	2030/A	2030/B	2050 BAU	2050/A	2050/B
pre-1991	50	0	0	0	0	0	0	0	0
1991-93	50	0	0	0	0	0	0	0	0
1994-97	130	130	0	0	0	0	0	0	0
1998-2003	160	185	160	60	0	0	0	0	0
2004-06	90	115	90	50	40	0	0	0	0
2007-09	80	110	195	50	40	50	0	0	0
2010+	440	460	532	840	900	811	1,000	750	500
PHEV	0	0	23	0	20	139	0	250	500
Total	**1,000**	**1,000**	**1,000**	**1,000**	**1,000**	**1,000**	**1,000**	**1,000**	**1,000**

An average emission factor for the fleet described by each scenario and for the BAU vehicle fleet was determined for each pollutant with the emissions factors (EFs) described above. The scenario RRF was then calculated as:

$$RRF = 1 - \text{Scenario EF/BAU EF} \qquad \text{Eq. C-1}$$

Note that use of emission factors to determine RRFs in this manner implies that other technical and operational parameters, such as engine load and power, are unchanged between the BAU and scenario analysis. Table C-4 shows the resulting Technological RRFs, applicable to a typical port.[115]

Table C-4. Emission Relative Reduction Factors for Drayage Technological Scenarios

Scenario	Overall Emission Reductions (%)							
	NOx	PM$_{2.5}$	VOC	BC	CO$_2$	Acetaldehyde	Benzene	Formaldehyde
2020/A	19%	43%	14%	43%	0%	10%	11%	7%
2020/B	48%	62%	35%	62%	1%	21%	25%	12%
2030/A	48%	34%	33%	34%	0%	20%	23%	14%
2030/B	60%	52%	39%	52%	4%	22%	26%	15%
2050/A	-	-	-	-	6%	-	-	-
2050/B	-	-	-	-	13%	-	-	-

C.2.2.2. Application of Emission Relative Reduction Factors

The typical port RRFs were applied to each port's BAU drayage truck emission inventory, determined under the BAU methodology. To approximate changes at the few ports where adjustments were made to the BAU drayage age distribution to account for local programs, the revised age distribution described by the scenario was compared to the local distribution. These ports have local programs in effect that would exceed the scenarios considered here. For those cases, no additional emission reductions from those included in the BAU case were realized. At all other ports, the RRFs described

---

[115] A "typical port" in this assessment is intended to establish a hypothetical port that that allows EPA to illustrate the relative impacts of a particular strategy and/or scenario.

above (Equation C-1) by pollutant and scenario were multiplied by each port's BAU emission inventory to estimate the emission reductions associated with each scenario. There was no change in age distribution in either 2030 or 2050 at all ports, so the full RRFs were applied.

C.3. Rail

C.3.1. Scenarios

As discussed in Section 5, the rail strategy scenarios were based on the future year BAU distributions of locomotives, which are shown in Tables C-5 and C-6.

Table C-5. BAU Distribution of Line-Haul Locomotives by Emissions Tier

Tier	2011	2020	2030	2050
Pre-Tier 0	10%	0%	0%	0%
Tier 0	37%	3%	0%	0%
Tier 0+	19%	33%	10%	0%
Tier 1	4%	0%	0%	0%
Tier 1+	6%	9%	5%	0%
Tier 2	24%	0%	0%	0%
Tier 2+	0%	22%	17%	0%
Tier 3	0%	10%	9%	0%
Tier 4	0%	23%	59%	100%
Total	**100%**	**100%**	**100%**	**100%**

Table C-6. BAU Distribution of Switcher Locomotives by Tier

Tier	2011	2020	2030	2050
Pre-Tier 0	74%	38%	8%	0%
Tier 0	7%	1%	0%	0%
Tier 0+	10%	45%	52%	0%
Tier 1	1%	0%	0%	0%
Tier 1+	0%	1%	1%	0%
Tier 2	7%	7%	0%	0%
Tier 2+	0%	0%	6%	0%
Tier 3	1%	3%	3%	0%
Tier 4	0%	5%	29%	100%
Total	**100%**	**100%**	**100%**	**100%**

Table C-7 shows the strategy scenarios that were analyzed as described in Section 6 and this appendix.

Table C-7. Rail Strategy Scenarios

Strategy	2020/A	2020/B	2030/A	2030/B	2050/A	2050/B
Line-haul—Technology Strategies	Replace 50% of Tier 0+ engines with Tier 2+ engines	Replace 100% of Tier 0+ engines with 50% 2+ engines and 50% Tier 4 engines	Replace 100% of Tier 1+ and earlier engines with 50% 2+ engines and 50% Tier 4 engines	Replace all pre-Tier engines with Tier 4 engines.	Replace 10% of Tier 4 with zero emission locomotive	Replace 25% of Tier 4 with zero emission locomotive
Line-haul—Operational Strategies	1% improvement in fuel efficiency	5% improvement in fuel efficiency	5% improvement in fuel efficiency	10% improvement in fuel efficiency	10% improvement in fuel efficiency	20% improvement in fuel efficiency
Switchers	Replace 50% of Pre-Tier 0 engines with 95% Tier 2+ engines and 5% Tier 4 Genset	Replace all Pre-Tier 0 engines with 90% Tier 2+ and 10% Tier 4 Genset	Replace all Pre-Tier 0 engines and 20% of Tier 0+ with 90% Tier 2+ engines and 10% Tier 4 Genset	Replace all Pre-Tier 0 engines and 40% of Tier 0+ with 70% Tier 4 engines and 30% Tier 4 Genset	Assume 30% Tier 4 Genset	Assume 50% Tier 4 Genset

C.3.2. Methodology and Assumptions

As described in Section 6, rail emission reductions were calculated by developing an RRF for each scenario strategy and pollutant, and these RRFs were calculated using average emission rates, determined as the emission rate for each engine tier weighted by the locomotive engine population distribution in that tier. An RRF was calculated from the following equation:

$$\text{RRF} = 1 - \text{Scenario EF/BAU EF} \qquad \text{Eq. C-2}$$

Table C-8 shows an example of how a line-haul locomotive RRF was estimated.

Table C-8. Example of Scenario 2020/A NOx Emission Reduction Factor for Line-Haul Locomotives

Tier	NOx Emission	Pop	Weighted	Pop	Weighted		Relative
Pre-Tier 0	13.0	0%	0.00	0%	0		
Tier 0	8.6	3%	0.27	3%	0.27		
Tier 0+	7.2	33%	2.38	17%	1.19		
Tier 1	6.7	0%	0.00	0%	0.00		
Tier 1+	6.7	9%	0.62	9%	0.62		
Tier 2	4.95	0%	0.00	0%	0.00		
Tier 2+	4.95	22%	1.08	38%	1.90		
Tier 3	4.95	10%	0.50	10%	0.50		
Tier 4	1.0	23%	0.23	23%	0.23		
		100%	5.08	100%	4.71		93%

In developing these strategies, it was assumed that any new engines would have similar duty cycles, rated power, and annual usage as the engines they replace, such that emission changes were due solely to changes in the engine emission rates (e.g., in g/kWh).

For line-haul locomotives, emission rates for NOx, $PM_{2.5}$, and HC for each line-haul tier were taken from EPA guidance.[116] VOC emission factors were calculated from hydrocarbon (THC) emission rates, using EPA conversion factors.[117] Select air toxics emission factors were computed from the VOC emissions factors using EPA speciation profiles for on-road engines from the MOVES model.[118] Well-to-wheels CO_2 emission factors (in g/ton-mi) were derived from GREET2015; the Rail emission factors for Freight Rails were used.[119] No change in fuel economy/CO_2 emission rates between locomotive engine tiers was assumed. Zero emissions locomotives were assumed to be electric locomotives with only well-to-plug CO_2 emissions. RRFs for BC were assumed to be equal to $PM_{2.5}$ RRFs.

For switcher locomotives, the RRFs were calculated using the emission rate for each tier weighted by the locomotive engine population distribution in that tier for the BAU emission inventory and strategy scenarios. While GenSet locomotives can be built to Tier 3 or Tier 4 standards, it was assumed that the Tier 4 standards were more appropriate for all GenSet locomotives in these scenarios. The RRFs were calculated as described by Equation C-2 above and the underlying methodology and assumptions were generally similar to the Line-haul Technology scenarios. For scenarios involving replacements with Tier 4 GenSets, it was assumed the GenSet engines would meet the EPA Nonroad Tier 4 emission standards. These were assumed to fall under the nonroad emission standards for engine power between 175 and 750 hp.

C.3.3. Application of Relative Reduction Factors

The RRF for each scenario and pollutant was multiplied by the relevant portion of the BAU emissions inventory for the appropriate analysis year, with the resulting line-haul and switcher locomotive emissions reductions for each scenario. This method was applied uniformly to all ports within this national scale analysis, depending upon the level of rail activity.

C.4. Cargo Handling Equipment

The analysis of emission reduction strategies for CHE focused on those equipment types that contribute the bulk of CHE emissions at most ports: yard tractors, rubber tire gantry (RTG) cranes, and container

---

[116] U.S. Environmental Protection Agency, *Technical Highlights: Emission Factors for Locomotives,* Report EPA-420-F-09-025, April 2009.

[117] U.S. Environmental Protection Agency, *Conversion Factors for Hydrocarbon Emission Components (July 2010)*, Report EPA-420-R-10-015 NR-002d.

[118] U.S. Environmental Protection Agency, *MOVES2010b: Additional Toxics Added to MOVES.* EPA-420-B-12-029a, May 2012, Sec 3.1.1. Available at: http://www.epa.gov/otaq/models/moves/documents/420b12029a.pdf.

[119] GREET2015 was released October 2, 2015. For information on the model, see https://greet.es.anl.gov/.

handlers (side picks and top picks, which are under the category of "rubber tire loaders" in the NEI). See Section 5 for further background on CHE strategies.

For each scenario, an RRF was calculated using the emission rate for each engine tier weighted by the population distribution in that tier. This approach was similar to the drayage and locomotive replacement strategy scenarios discussed above, such as in Table C-8 for Line-haul Locomotive Technology scenarios.

C.4.1. Yard Truck Strategy Scenarios

As discussed in Section 5, the yard truck scenarios were based on the NONROAD model's future year distributions of yard trucks, as shown in Table C-9.

Table C-9. Distribution of Yard Trucks by Tier

Tier	2011	2020	2030	2050
Tier 1	9%	0%	0%	0%
Tier 2	17%	0%	0%	0%
Tier 3	64%	3%	0%	0%
Tier 4	10%	97%	100%	100%
Total	100%	100%	100%	100%

Table C-10 shows the BAU assumptions that are relevant for all yard truck strategy scenarios, since most, if not all, yard trucks are already assumed to be meeting Tier 4 emission standards in all analysis years, based on the assumptions for this assessment. Thus, all strategy scenarios focus on the introduction of battery electric yard truck technologies.

Table C-10. Yard Truck Strategy Scenarios

2020/A	2020/B	2030/A	2030/B	2050/A	2050/B
Replace all Tier 3 with Tier 4	Replace all Tier 3 with Tier 4, and replace 5% of Tier 4 with battery electric	Replace 10% Tier 4 diesel with battery electric	Replace 25% Tier 4 diesel with battery electric	Replace 25% of Tier 4 diesel engines with battery electric	Replace 50% of Tier 4 diesel engines with battery electric

C.4.2. Cranes Strategy Scenarios

As discussed in Section 5, the crane strategy scenarios were also based on the future year distributions of cranes, as assumed in the NONROAD model and shown in Table C-11.

Table C-11. Distribution of RTG Cranes by Tier

Tier	2011	2020	2030	2050
Uncontrolled	6%	1%	0%	0%
Tier 1	27%	3%	0%	0%
Tier 2	20%	5%	1%	0%
Tier 3	38%	17%	2%	0%
Tier 4	9%	74%	98%	100%
Total	100%	100%	100%	100%

Table C-12 shows the RTG crane strategy scenarios that were analyzed, including the significant increase of electric crane technologies for most analysis years. Since this technology is available today and is being installed at a number of leading ports, this assessment assumed that electric cranes would be widely deployed by 2050.

Table C-12. RTG Crane Strategy Scenarios

2020/A	2020/B	2030/A	2030/B	2050/A	2050/B
Replace all Uncontrolled and 50% of Tier 1 and 2 with 50% Tier 3 and 50% Tier 4	Replace all Uncontrolled, Tier 1 and 2 with 75% Tier 4 and 25% electric	Replace all Tier 2 and 3 with 50% Tier 4 and 50% electric. Replace 10% Tier 4 with electric	Replace all Tier 2 and 3 with 50% Tier 4 and 50% electric. Replace 25% Tier 4 with electric	Replace 50% Tier 4 with electric	Replace 75% Tier 4 with electric

C.4.3. Container Handler Strategy Scenarios

The container handler strategy scenarios were based on the future year distributions shown in Table C-13.

Table C-13. Distribution of Container Handlers by Tier

Tier	2011	2020	2030	2050
Uncontrolled	2%	0%	0%	0%
Tier 1	26%	1%	0%	0%
Tier 2	23%	2%	0%	0%
Tier 3	44%	15%	0%	0%
Tier 4	5%	81%	100%	100%
Total	100%	100%	100%	100%

Table C-14 shows the container handler strategy scenarios used in this assessment, as described further in Section 6.

Table C-14. Container Handler Strategy Scenarios

2020/A	2020/B	2030/A	2030/B	2050/A	2050/B
Replace all Tier 1 and 2 engines with 50% Tier 3 and 50% Tier 4.	Replace Tier 1 and 2 engines with Tier 4 engines. Replace Tier 3 with 50% Tier 4 and 50% elec. engines	Replace 10% of Tier 4 diesel engines with electric engines	Replace 25% of Tier 4 diesel engines with electric	Replace 50% of Tier 4 diesel engines with electric	Replace 75% of Tier 4 diesel engines with electric

C.4.4. Relative Reduction Factors: Methodology and Assumptions for all CHE

For each scenario, an RRF was calculated using the emission rate for each engine tier weighted by the population distribution in that tier.[120] EPA's emission standards were used as criteria pollutant emission factors for conventional diesel equipment.[121] Nonroad emission standards vary based on the rated power of the engine:[122]

- Yard trucks were assumed to fall in the rated power category of 130 to 225 kW (175 to 300 hp) based upon an average engine size of 206 hp.

- RTG cranes were assumed to fall in the rated power category of 225 to 450 kW (300-600 hp) based on an average engine size of 453 hp.

- Container handlers were assumed within the rated power category of 130 to 225 kW (175 to 300 hp) based on an average engine size of 184 hp for side handlers and 282 hp for top handlers.

Emission factors for pre-Tier engines were taken from the baseline CHE emission factors used in this assessment.

VOC emission rates were based off of the hydrocarbon (THC) standards, using EPA conversion factors.[123] Select air toxic emission factors were computed from the VOC emission factors using EPA speciation profiles for on-road engines from the MOVES model.[124] Diesel upstream wheel-to-pump CO_2 emission factors were calculated using GREET 2015. CHE tailpipe CO_2 emission factors were obtained from EPA's

---

[120] Note that the baseline and BAU emission inventories for CHE include a mix of fuels. However, the strategy scenarios that were analyzed imply replacement of diesel engines only. Thus, RRF calculations were made in terms of emission factors for diesel engines only. However, according to the NONROAD model used, the three types of equipment considered here are predominately diesel fueled for all years included here, so any bias in the results from this assumption was minimal.

[121] U.S. Environmental Protection Agency, *Nonroad Compression-Ignition Emission Standards*. Available at: http://www.epa.gov/otaq/standards/nonroad/nonroadci.htm.

[122] U.S. Environmental Protection Agency, *Current Methodologies in Preparing Mobile Source Port-Related Emission Inventories*, April 2009.

[123] U.S. Environmental Protection Agency, *Conversion Factors for Hydrocarbon Emission Components (July 2010)*, Report EPA-420-R-10-015 NR-002d.

[124] U.S. Environmental Protection Agency, *MOVES2010b: Additional Toxics Added to MOVES*. EPA-420-B-12-029a, May 2012, Sec 3.1.1. Available at: http://www.epa.gov/otaq/models/moves/documents/420b12029a.pdf.

Current Methodologies document. For electric equipment, no tailpipe pollutant emissions were assumed. Upstream electricity emission factors for CO_2 were calculated using GREET2015 assuming an average U.S. grid mix for the appropriate scenario years. Well-to-wheels emission factors assumed the sum of upstream and downstream emissions. RRFs for BC were taken as equal to those for $PM_{2.5}$.

This method was applied uniformly to all ports in this national scale analysis. However, the calculated reductions were based on the resolution of the BAU inventories. As with other CHE types, the calculated reductions are likely to overestimate the potential percent reduction in emissions at ports that have CHE that is newer than average, while it will underestimate the reductions at ports with older equipment.

C.4.5. Application of Relative Reduction Factors

Generally, the RRF for each strategy scenario was multiplied by the CHE portion of the BAU emissions inventory for the appropriate analysis year to determine the emission reductions for each scenario.

In the case of CHE, as described above, the BAU inventory did not include resolution by equipment type or operating mode with only total CHE emissions presented for each pollutant for each port. This method was applied uniformly.

C.5. Harbor Craft

The analysis of emission reduction strategies for harbor craft focused on the two types of vessels that contribute the bulk of harbor craft emissions at most ports: tugs and ferries. See Section 5 for further background on harbor craft strategies.

C.5.1. Scenarios: Tugs

As described in Section 5, the future year distribution of tugs was estimated using a methodology based on the growth and scrappage assumptions in EPA's NONROAD model, as shown in Table C-15.

Table C-15. Distribution of Tugs by Tier

Tier	2011	2020	2030	2050
Tier 0/0+	61%	10%	0%	0%
Tier 1/1+	35%	24%	3%	0%
Tier 2/2+	4%	33%	7%	0%
Tier 3/3+	0%	30%	80%	61%
Tier 4	0%	3%	10%	39%
Total	**100%**	**100%**	**100%**	**100%**

Table C-16 shows the tug strategy scenarios that were analyzed in this assessment. Due to the slower national distribution of fleet turnover for tugs in the NONROAD model, the strategies analyzed included more repowers and replacements to cleaner diesel engines for all analysis years, with some opportunity for hybrid electric technology in 2030 and 2050.

Table C-16. Tug Strategy Scenarios

2020/A	2020/B	2030/A	2030/B	2050/A	2050/B
Repower/ Replace all Pre-Control engines with Tier 3 engines	Repower/ Replace all Pre-Control and Tier 1 with Tier 3 Repower 10% of Tier 2 with Tier 3 hybrid electric	Repower/ Replace all Tier 1 and 2 with Tier 4 Repower/ Replace 25% of Tier 3 engines with Tier 4 engines	Repower/ Replace all Tier 1 and 2 with Tier 4 Repower/ Replace 50% of Tier 3 engines with Tier 4 engines Repower/ Replace 25% of Tier 4 with hybrid electric	Repower/ Replace 50% of Tier 3 engines with Tier 4 engines Repower/ Replace 10% of Tier 4 with hybrid electric	Repower/ Replace all Tier 3 engines with Tier 4 engines Repower/ Replace 25% of Tier 4 with hybrid electric

C.5.2. Scenarios: Ferries

As with tugs, the future year distributions of ferries were estimated using a methodology based on the growth and scrappage assumptions in EPA's NONROAD model. Table C-17 shows that distribution.

Table C-17. Distribution of Ferries by Tier

Tier	2011	2020	2030	2050
Tier 0/0+	75%	39%	10%	0%
Tier 1/1+	21%	18%	12%	0%
Tier 2/2+	4%	10%	8%	1%
Tier 3/3+	0%	28%	59%	60%
Tier 4	0%	5%	11%	39%
Total	100%	100%	100%	100%

Table C-18 shows the ferry strategy scenarios. As with tugs, the slower fleet turnover assumed for ferries in the BAU inventory allowed for cleaner diesel strategies to be analyzed in all analysis years, in addition to hybrid electric technology in 2030 and 2050.

Table C-18. Ferry Strategy Scenarios

2020/A	2020/B	2030/A	2030/B	2050/A	2050/B
Repower/ Replace all Pre-Control engines with Tier 3 engines	Repower/ Replace all Pre-Control and Tier 1 with Tier 3 Repower 10% of Tier 2 with Tier 3 hybrid electric	Repower/ Replace all Tier 0, 1 and 2 with Tier 4 Repower/ Replace 25% of Tier 3 engines with Tier 4 engines	Repower/ Replace all Tier 0, 1 and 2 with Tier 4 Repower/ Replace 50% of Tier 3 engines with Tier 4 engines. Repower/ Replace 25% of Tier 4 with hybrid electric	Repower/ Replace all Tier 2 and 50% of Tier 3 engines with Tier 4 engines Repower/ Replace 10% of Tier 4 with hybrid electric	Repower/ Replace all Tier 2 and 3 engines with Tier 4 engines Repower/ Replace 25% of Tier 4 with hybrid electric

C.5.3. Development and Application of Relative Reduction Factors

For each scenario, an RRF was developed using the emission rate for each engine tier weighted by the population distribution in that tier. This process is similar to the example for line-haul locomotives discussed earlier in this appendix. The RRF for each scenario was then multiplied by the BAU emissions for the given vessel type appropriate analysis year to determine the scenario's emission reductions.

The following paragraphs further describe how this was done in conjunction with more complex methodology for the harbor craft BAU inventory development.

To determine the BAU emission inventory by vessel type, the relative share of emissions by each vessel type (tugs or ferries), by pollutant, within the two vessel categories (goods moving or non-goods moving) was determined for the baseline year (e.g., tugs average 97% of CO emissions from goods moving vessels nationwide). This share was then assumed to also apply for future years, which was reasonable since the vessel categories are grown rather than individual vessel types. Next, the share of emissions by operating mode that were due to vessels in each category was determined. For example, the product of the tug share of total goods moving emissions by pollutant, along with the share of total emissions due to goods movements by operating mode, approximated the BAU tug inventory by pollutant by operating mode in each scenario analysis year. A similar approach was applied for ferries.

Finally, the emission reduction from the application of each strategy was determined as the product of the vessel and mode-specific BAU inventory and the RRF determined for each strategy scenario. This method was applied consistently for all strategy scenarios and applied uniformly.

C.6. Ocean Going Vessels

As discussed in Section 6, OGV strategies were grouped in the following scenario categories:

- Fuel Changes
- Shore Power
- Stack Bonnets
- Reduced Hoteling Time

Reductions were calculated for all scenarios relative to the BAU inventories, independently. This method considered reductions separately for each scenario, consistent with other sectors. Note that in practice results may not be additive as such relative to the BAU case.[125] In summary, the reductions presented here would be reasonable for each individual strategy, but would overestimate the cumulative impact if multiple strategies were applied simultaneously.

---

[125] For example, if shore power was applied after fuel changes were required, the BAU inventory that would exist in practice for shore power would be smaller for many pollutants, as would be the reductions.

C.6.1. Fuel Changes Scenarios

The Fuel Change strategy scenarios are found in Table C-19 and Table C-20 for propulsion and auxiliary engines, respectively.

Table C-19. Fuel Change Strategy Scenarios for OGV Propulsion Engines

Ship Type	2020/A	2020/B	2030/A	2030/B	2050/A	2050/B
Bulk	10% use 500 ppm sulfur fuel; 2% use LNG	25% use 500 ppm sulfur fuel; 10% use LNG	25% use 200 ppm sulfur fuel; 4% use LNG	50% use 200 ppm sulfur fuel; 15% use LNG	8% use LNG	25% use LNG
Container	10% use 500 ppm sulfur fuel; 1% use LNG	25% use 500 ppm sulfur fuel; 5% use LNG	25% use 200 ppm sulfur fuel; 2% use LNG	50% use 200 ppm sulfur fuel; 5% use LNG	5% use LNG	5% use LNG
Passenger	10% use 500 ppm sulfur fuel	25% use 500 ppm sulfur fuel	25% use 200 ppm sulfur fuel	50% use 200 ppm sulfur fuel	-	-
Tanker	10% use 500 ppm sulfur fuel; 2% use LNG	25% use 500 ppm sulfur fuel; 10% use LNG	25% use 200 ppm sulfur fuel; 4% use LNG	50% use 200 ppm sulfur fuel; 15% use LNG	8% use LNG	25% use LNG

Table C-20. Fuel Change Scenarios for OGV Auxiliary Engines

Ship Type	2020/A	2020/B	2030/A	2030/B	2050/A	2050/B
Bulk	10% use ULSD; 2% use LNG	20% use ULSD; 10% use LNG	30% use ULSD; 4% use LNG	40% use ULSD; 15% use LNG	8% use LNG	25% use LNG
Container	10% use ULSD; 1% use LNG	20% use ULSD; 5% use LNG	30% use ULSD; 2% use LNG	40% use ULSD; 5% use LNG	5% use LNG	5% use LNG
Passenger	10% use ULSD	20% use ULSD	30% use ULSD	40% use ULSD	-	-
Tanker	10% use ULSD; 2% use LNG	20% use ULSD; 10% use LNG	30% use ULSD; 4% use LNG	40% use ULSD; 15% use LNG	8% use LNG	25% use LNG

LNG is limited in container ships to 5% based upon a study by Lloyds,[126] and no LNG use was included for passenger ships due to passenger safety issues.

---

[126] Lloyds Register Marine, *Global Marine Fuel Trends 2030,* 2014.

C.6.1.1. Developing and Application of Relative Reduction Factors

RRFs were calculated from the ratio of average emission factors under the BAU scenario and those under each analysis scenario. RRFs are calculated according to the Equation C-2.

To apply these scenarios to the 2020, 2030, and 2050 BAU emissions inventories, emission factors were developed by engine type and fuel type. Table C-21 and Table C-22 show these emission factors for propulsion and auxiliary engines.

Table C-21. Average Emission Factors for Propulsion Engines

Engine Type	Sulfur	Emission Factors (g/kWh)					
		PM$_{2.5}$	HC	BC	CO$_2$	SO$_2$	BSFC
MSD	0.10%	0.17	0.5	0.0102	646	0.4	203
	0.05%	0.16	0.5	0.0093	646	0.2	203
	0.02%	0.15	0.5	0.0088	646	0.08	203
SSD	0.10%	0.17	0.6	0.0104	589	0.36	185
	0.05%	0.16	0.6	0.0096	589	0.18	185
	0.02%	0.15	0.6	0.0092	589	0.07	185
GT	0.10%	0.25	0.1	0.0151	923	0.57	290
	0.05%	0.23	0.1	0.0138	923	0.28	290
	0.02%	0.22	0.1	0.0131	923	0.11	290
Otto	LNG	0.03	0.5	0.0105	457	0.003	166

Table C-22. Average Emission Factors for Auxiliary Engines

Sulfur	Emission Factors (g/kWh)					
	PM$_{2.5}$	HC	BC	CO$_2$	SO$_2$	BSFC
0.10%	0.17	0.4	0.0101	691	0.42	217
ULSD	0.14	0.4	0.0082	691	0.01	217
LNG	0.03	0.5	0.0105	457	0.003	166

Here, the brake specific fuel consumption (BSFC) for medium speed diesel (MSD), slow speed diesel (SSD) and gas turbines (GT) was taken from a European Union study (Entec).[127] Emission factors and BSFC for LNG were taken from an IMO study.[128] As discussed in Appendix A, PM$_{10}$ was calculated using the formula:

$$\text{PM}_{10} \text{ EF} = 0.23 + \text{BSFC} \times 7 \times 0.02247 \times (\text{Fuel Sulfur Fraction} - 0.0024) \qquad \text{Eq. C-3}$$

---

[127] Entec UK Limited, *Quantification of Emissions from Ships Associated with Ship Movements between Ports in the European Community*, prepared for the European Commission, July 2002.

[128] International Maritime Organization, *Third IMO GHG Study*, June 2014. Available at: http://www.imo.org/en/OurWork/Environment/PollutionPrevention/AirPollution/Pages/Relevant-links-to-Third-IMO-GHG-Study-2014.aspx.

$PM_{2.5}$ was taken as 92% of PM_{10}. SO_2 was calculated using the formula:

$$SO_2 \text{ EF} = BSFC \times 2 \times 0.97753 \times \text{Fuel Sulfur Fraction} \qquad \text{Eq. C-4}$$

CO_2 emission factors were calculated as follows for diesel:

$$CO_2 \text{ EF} = BSFC \times 0.868 \times 3.667 \qquad \text{Eq. C-5}$$

CO_2 emission factors were calculated as follows for LNG: [129]

$$CO_2 \text{ EF} = BSFC \times 2.75 \qquad \text{Eq. C-6}$$

BC was calculated as 6% of $PM_{2.5}$ for diesel fuels and 38% of $PM_{2.5}$ for LNG based upon the EPA's Black Carbon Report to Congress.[130]

Note that, consistent with the baseline and BAU emission inventory development, OGV uses total hydrocarbons (HC) while all other sectors report volatile organic compounds (VOC). HC emission factors for SSD, MSD, GT and ST engines came from Entec.

Average NOx emission factors by engine type for 2020 and 2030 were calculated as discussed in Appendix B. They are listed by calendar year and engine type in Table C-23 for propulsion engines and by calendar year and ship type in Table C-24 for auxiliary engines. LNG NOx emission factors were taken from the IMO study.

Table C-23. Average Propulsion NOx Emission Factors

Engine Type	NOx (g/kWh)	
	2020	2030
MSD	9.4	3.7
SSD	10.6	5.0
GT	5.7	5.7
LNG Otto	1.3	1.3

Table C-24. Average Auxiliary NOx Emission Factors

Engine Type	NOx (g/kWh)	
	2020	2030
Passenger	10.3	3.7
Other	8.6	4.1
LNG Ships	1.3	1.3

---

[129] International Maritime Organization, *Third IMO GHG Study*, June 2014.

[130] U.S. Environmental Protection Agency, *Report to Congress on Black Carbon,* EPA-450/D-11-001, March 2011.

To calculate aggregate emission factors for the BAU and fuel scenarios, engine-type weighting factors were needed. Using the combined Entrances and Clearances data combined with Lloyd's data for the ports, weighting factors were calculated by ship type. Table C-25 shows these factors.[131]

Table C-25. Engine Type Weighting Factors by Ship Type

Ship Type	MSD	SSD	GT
Bulk Carrier	0%	100%	0%
Container	3%	97%	0%
Passenger	89%	0%	11%
Tanker	3%	97%	0%

To calculate emission reductions for the scenarios listed in Table C-19 and Table C-20, BAU emission inventories were separated into propulsion and auxiliary engine emissions for the four ship types. Emissions related to propulsion engines during reduced speed zone (RSZ) and maneuvering modes were combined into the propulsion engine emissions. Emissions related to auxiliary engines during RSZ, maneuvering and hoteling modes were combined into the auxiliary engine emissions. In addition, hoteling-only emissions (from auxiliary engines) were also calculated by ship type to use for strategies that these emissions.

Table C-26 presents average emissions factors by ship type calculated from the BAU emission inventories (i.e., "BAU EF" in Equation C-2). BAU emissions assume the use of 1,000 ppm S MDO/MGO in both propulsion and auxiliary engines; BAU CO_2 emission factors stay constant through 2050. Table C-27 through Table C-31 show the average emission factors by ship type for the scenarios (i.e., "Scenario EF" in Equation C-2). See Section 6 for the RRFs that were calculated for each scenario.

Table C-26. BAU Average Emission Factors by Ship Type and Calendar Year

Engine	Ship Type	CY	Combined Emission Factors (g/kWh)					
			NOx	PM2.5	HC	BC	CO2	SO2
Propulsion	Bulk	2020	10.6	0.17	0.6	0.010	589	0.36
		2030	5.0	0.17	0.6	0.010	589	0.36
	Container	2020	10.6	0.17	0.6	0.010	591	0.36
		2030	4.9	0.17	0.6	0.010	591	0.36
	Passenger	2020	9.0	0.18	0.5	0.011	677	0.42
		2030	3.9	0.18	0.5	0.011	677	0.42
	Tanker	2020	10.6	0.17	0.6	0.010	591	0.36
		2030	4.9	0.17	0.6	0.010	591	0.36
Auxiliary	Bulk	2020	8.6	0.17	0.4	0.010	691	0.42
		2030	4.1	0.17	0.4	0.010	691	0.42
	Container	2020	8.6	0.17	0.4	0.010	691	0.42
		2030	4.1	0.17	0.4	0.010	691	0.42
	Passenger	2020	10.3	0.17	0.4	0.010	691	0.42
		2030	3.7	0.17	0.4	0.010	691	0.42
	Tanker	2020	8.6	0.17	0.4	0.010	691	0.42
		2030	4.1	0.17	0.4	0.010	691	0.42

---

[131] Steam turbines (ST) were not significant and included here.

Table C-27. Average Emission Factors by Ship Type for Fuel Scenario 2020/A

Engine	Vessel	Combined Emission Factors (g/kWh)					
		NOx	PM$_{2.5}$	HC	BC	CO$_2$	SO$_2$
Propulsion	Bulk	10.4	0.17	0.6	0.010	586	0.34
	Container	10.5	0.17	0.6	0.010	589	0.34
	Passenger	9.0	0.18	0.5	0.011	677	0.39
	Tanker	10.4	0.17	0.6	0.010	588	0.34
Auxiliary	Bulk	8.5	0.16	0.4	0.010	686	0.37
	Container	8.6	0.16	0.4	0.010	688	0.38
	Passenger	10.3	0.16	0.4	0.010	691	0.38
	Tanker	8.5	0.16	0.4	0.010	686	0.37

Table C-28. Average Emission Factors by Ship Type for Fuel Scenario 2020/B

Engine	Vessel	Combined Emission Factors (g/kWh)					
		NOx	PM$_{2.5}$	HC	BC	CO$_2$	SO$_2$
Propulsion	Bulk	9.7	0.16	0.6	0.010	576	0.28
	Container	10.1	0.16	0.6	0.010	584	0.3
	Passenger	9.0	0.18	0.5	0.011	677	0.36
	Tanker	9.7	0.16	0.6	0.010	577	0.28
Auxiliary	Bulk	7.9	0.15	0.4	0.010	667	0.3
	Container	8.3	0.15	0.4	0.010	679	0.32
	Passenger	10.3	0.16	0.4	0.010	691	0.32
	Tanker	7.9	0.15	0.4	0.010	667	0.3

Table C-29. Average Emission Factors by Ship Type for Fuel Scenario 2030/A

Engine	Vessel	Combined Emission Factors (g/kWh)					
		NOx	PM$_{2.5}$	HC	BC	CO$_2$	SO$_2$
Propulsion	Bulk	4.8	0.16	0.6	0.010	584	0.28
	Container	4.9	0.17	0.6	0.010	588	0.28
	Passenger	3.9	0.17	0.5	0.010	677	0.33
	Tanker	4.8	0.16	0.6	0.010	585	0.28
Auxiliary	Bulk	4.0	0.15	0.4	0.010	681	0.28
	Container	4.0	0.16	0.4	0.010	686	0.29
	Passenger	3.7	0.16	0.4	0.010	691	0.3
	Tanker	4.0	0.15	0.4	0.010	681	0.28

Table C-30. Average Emission Factors by Ship Type for Fuel Scenario 2030/B (g/kWh)

Engine	Vessel	Combined Emission Factors (g/kWh)					
		NOx	PM$_{2.5}$	HC	BC	CO$_2$	SO$_2$
Propulsion	Bulk	4.4	0.14	0.6	0.010	569	0.16
	Container	4.7	0.16	0.6	0.010	584	0.2
	Passenger	3.9	0.17	0.5	0.010	677	0.25
	Tanker	4.4	0.14	0.6	0.010	570	0.16
Auxiliary	Bulk	3.7	0.13	0.4	0.009	656	0.19
	Container	3.9	0.15	0.4	0.009	679	0.24
	Passenger	3.7	0.16	0.4	0.009	691	0.26
	Tanker	3.7	0.13	0.4	0.009	656	0.19

Table C-31. Average CO$_2$ Emission Factors for Fuel Scenarios 2050/A and 2050/B

Engine	Vessel	Scenario EFs (g/kWh)	
		2050/A	2050/B
Propulsion	Bulk	578	556
	Container	584	584
	Passenger	677	677
	Tanker	580	557
Auxiliary	Bulk	672	632
	Container	679	679
	Passenger	691	691
	Tanker	672	632

C.6.2. Shore Power Scenarios

Table C-32 presents the Shore Power strategy scenarios for container, passenger and reefer ships that stop at the ports that were part of this national scale analysis.[132]

Table C-32. Shore Power Strategy Scenarios

Ship Type	2020/A	2020/B	2030/A	2030/B	2050/A	2050/B
Container	1%	10%	5%	20%	15%	35%
Passenger	10%	20%	20%	40%	30%	60%
Reefer	1%	5%	5%	10%	10%	20%

Table C-32 shows the penetration rate for the three ship types for each scenario, defined in terms of installed auxiliary power. To correspond with CARB's shore power regulation,[133] shore power was applied to container, passenger and reefer ships in this assessment.

---

[132]Shore power was not applied to RoRos as hoteling emissions from RoRos was much smaller than the other three ship types.
[133] CARB, *Airborne Toxic Control Measure for Auxiliary Diesel Engines Operated on Ocean-Going Vessels At- Berth in a California Port,* Final Regulation Order, 2010. Available at: http://www.arb.ca.gov/ports/shorepower/finalregulation.pdf.

C.6.2.1. Defining Frequent Callers

Percentages in Table C-32 represent the percentage of installed auxiliary power that shore power is applied to for frequent by ship type at each port from the 2011 Entrances and Clearances data.[134] Installed power directly related to emissions for a given ship type, so by specifying the percent of installed power related to frequent callers, the amount of eligible frequent caller emissions was estimated.

Table C-33 shows the resulting percentages due to frequent callers by port and ship type. Port-vessel combinations marked N/A had no ships of that ship type stop at the port.

Table C-33. Frequent Caller[135] Percentages by Port and Ship Type

Port	Ship Type	% Frequent Caller	Port	Ship Type	% Frequent Caller
New York / New Jersey	Container	62%	Baltimore	Container	31%
	Passenger	93%		Passenger	97%
	Reefer	87%		Reefer	N/A
New Orleans	Container	57%	Norfolk	Container	48%
	Passenger	99%		Passenger	84%
	Reefer	0%		Reefer	N/A
Miami	Container	62%	Philadelphia	Container	40%
	Passenger	98%		Passenger	N/A
	Reefer	83%		Reefer	56%
South Louisiana	Container	0%	Charleston	Container	57%
	Passenger	N/A		Passenger	83%
	Reefer	N/A		Reefer	0%
Seattle	Container	65%	Corpus Christi	Container	N/A
	Passenger	97%		Passenger	N/A
	Reefer	N/A		Reefer	N/A
Baton Rouge	Container	0%	Tampa	Container	72%
	Passenger	N/A		Passenger	100%
	Reefer	N/A		Reefer	0%
Port Arthur	Container	N/A	Savannah	Container	53%
	Passenger	N/A		Passenger	0%
	Reefer	N/A		Reefer	N/A
Portland	Container	75%	Coos Bay	Container	N/A
	Passenger	N/A		Passenger	N/A
	Reefer	N/A		Reefer	N/A
Mobile	Container	41%	San Juan	Container	75%
	Passenger	97%		Passenger	93%
	Reefer	0%		Reefer	95%
Houston	Container	61%	Port Average	Container	56%
	Passenger	0%		Passenger	96%
	Reefer	0%		Reefer	72%

---

[134] U.S. Army Corps of Engineers, *Vessel Entrances and Clearances*. Available at: http://www.navigationdatacenter.us/data/dataclen.htm.

[135] Frequent callers were defined for this assessment as individual vessels calling at a port 6 times or more times per year, and for passenger ships, 5 calls or more per year.

C.6.2.2. Relative Reduction Factor

Emission reductions for each ship type at the port are calculated as the BAU emissions times the RRF where RRF is defined as:

$$RRF = FC \times PR \times Eff \qquad \text{Eq. C-7}$$

Where

RRF is the relative reduction factor,

FC is the percent of installed power for frequent callers,

PR is the technology penetration levels, and

Eff is the emission reduction effectiveness (shore power emission reduction per call).

The assessment assumed approximately 2 hours to connect and disconnect cables during a call, and the strategy's effectiveness was based upon the number of hours connected versus the total hoteling time. Average hoteling times by vessel type were used to calculate effectiveness by ship type, and then the resulting valued to all ports with applicable vessel types. The same share of installed power by ship type by port was also applied for all future years. Shore power effectiveness was based on the number of hours connected divided by total average hoteling time. The number of hours connected was calculated as the total average hoteling time minus 2 hours.

Table C-34 shows per call effectiveness for shore power by ship type, considering only emissions from the vessels themselves.

Table C-34. Shore Power Effectiveness for Vessel Emissions Only, per call

Ship Type	Average Hoteling Time (hrs)	Shore Power Reduction (%)
Container	30.7	93%
Passenger	10.1	80%
Reefer	64.3	97%

In addition to vessel emissions, CO_2 and criteria pollutant emissions were assumed to be generated by power plants generating electricity for the shore power technology. Based upon the default U.S. average generation mix[136] using the Argonne National Laboratory's GREET 1 2015 model[137], emission

---

[136] The CO_2 emission rates for this assessment were calculated based on GREET, assuming an average U.S. generation mix. In practice, the generation mix could be significantly different. For example, in the Northwest, much of the electricity comes from hydropower, and therefore, those utilities emit less CO_2 overall.

[137] Argonne National Laboratories, *GREET Model 2015,* https://greet.es.anl.gov/.

factors for electricity generation[138,139] are shown in Table C-35. These should be compared to the auxiliary engine emission factors to assess the net effectiveness of shore power, considering the time spent plugged in for a given call.

Table C-35. Power Plant Emission Factors at Plug (g/kWh)

Year	NOx	PM10	PM2.5	HC	BC	CO2	SO2
2020	0.119	0.037	0.015	0.004	0.001	489	0.67
2030	0.124	0.040	0.016	0.005	0.001	478	0.633
2050	-	-	-	-	-	460	-

Similar to the special cases defined for drayage vehicles, the BAU emissions accounted for the use of shore power currently planned, as described in the BAU methodology.[140] BAU emissions for the two ports were modified to separate out power plant emissions from auxiliary engine emissions for those ships that use shore power and power plant criteria pollutant emissions were added. See Section 6 for more information on the resulting RRFs and Shore Power strategy scenario results.

C.6.3. Advanced Marine Emission Control System Scenarios

Advanced Marine Emission Control Systems (AMECS, and sometimes referred to as "stack bonnets") can also provide emission reductions while a ship is at berth. Table C-36 shows the penetration rates for the qualifying vessels for each AMECS strategy scenario analyzed in this assessment. Again, the percentages represent installed auxiliary power as a surrogate for hoteling emissions.

Table C-36. AMECS Strategy Scenarios

Ship Type	2020/A	2020/B	2030/A	2030/B
Container	1%	5%	5%	10%
Tanker	1%	5%	5%	10%

As discussed in Section 6, the AMECS strategy scenarios were targeted to only non-frequent callers for container and tanker ship types. Table C-37 shows the percentages due to non-frequent callers by port and ship type in this national scale analysis. Port-vessel combinations marked N/A had no ships of that ship type stop at the port. Overall, non-frequent callers were 47%.

---

[138] GREET2015 only extends to 2040, so the 2040 CO2 emission rate for power plants was used for 2050.

[139] Note that cargo loading and unloading occur while the connection is being made and removed, so the total hoteling time estimate is expected to be unchanged by shore power, although on a first call at a new terminal commissioning is required, which takes much longer. This calculation assumes 2 hours per call for connection and disconnection, but does not include any commissioning time.

[140] Only the Ports of Seattle and New York/New Jersey had installed shore power at the time of this assessment, and only for passenger ships.

Table C-37. Non-Frequent Caller Percentages by Port and Ship Type

Port	Ship Type	% Non-Frequent Caller	Port	Ship Type	% Non-Frequent Caller
New York / New Jersey	Container	38%	Baltimore	Container	69%
	Tanker	83%		Tanker	100%
New Orleans	Container	43%	Norfolk	Container	52%
	Tanker	97%		Tanker	100%
Miami	Container	38%	Philadelphia	Container	60%
	Tanker	100%		Tanker	87%
South Louisiana	Container	100%	Charleston	Container	43%
	Tanker	94%		Tanker	95%
Seattle	Container	35%	Corpus Christi	Container	N/A
	Tanker	100%		Tanker	87%
Baton Rouge	Container	100%	Tampa	Container	28%
	Tanker	87%		Tanker	66%
Port Arthur	Container	N/A	Savannah	Container	47%
	Tanker	61%		Tanker	100%
Portland	Container	25%	Coos Bay	Container	N/A
	Tanker	100%		Tanker	N/A
Mobile	Container	59%	San Juan	Container	25%
	Tanker	95%		Tanker	50%
Houston	Container	39%	Port Average	Container	44%
	Tanker	70%		Tanker	81%

See Section 6 for further details on the AMECS strategy scenarios and results.

C.7. Sector-by-sector Review of Results

C.7.1. Absolute Emission Reductions

The following figures show the absolute emission reductions obtained from applying each strategy as described above.

Appendix C: Analysis of Emission Reduction Scenarios

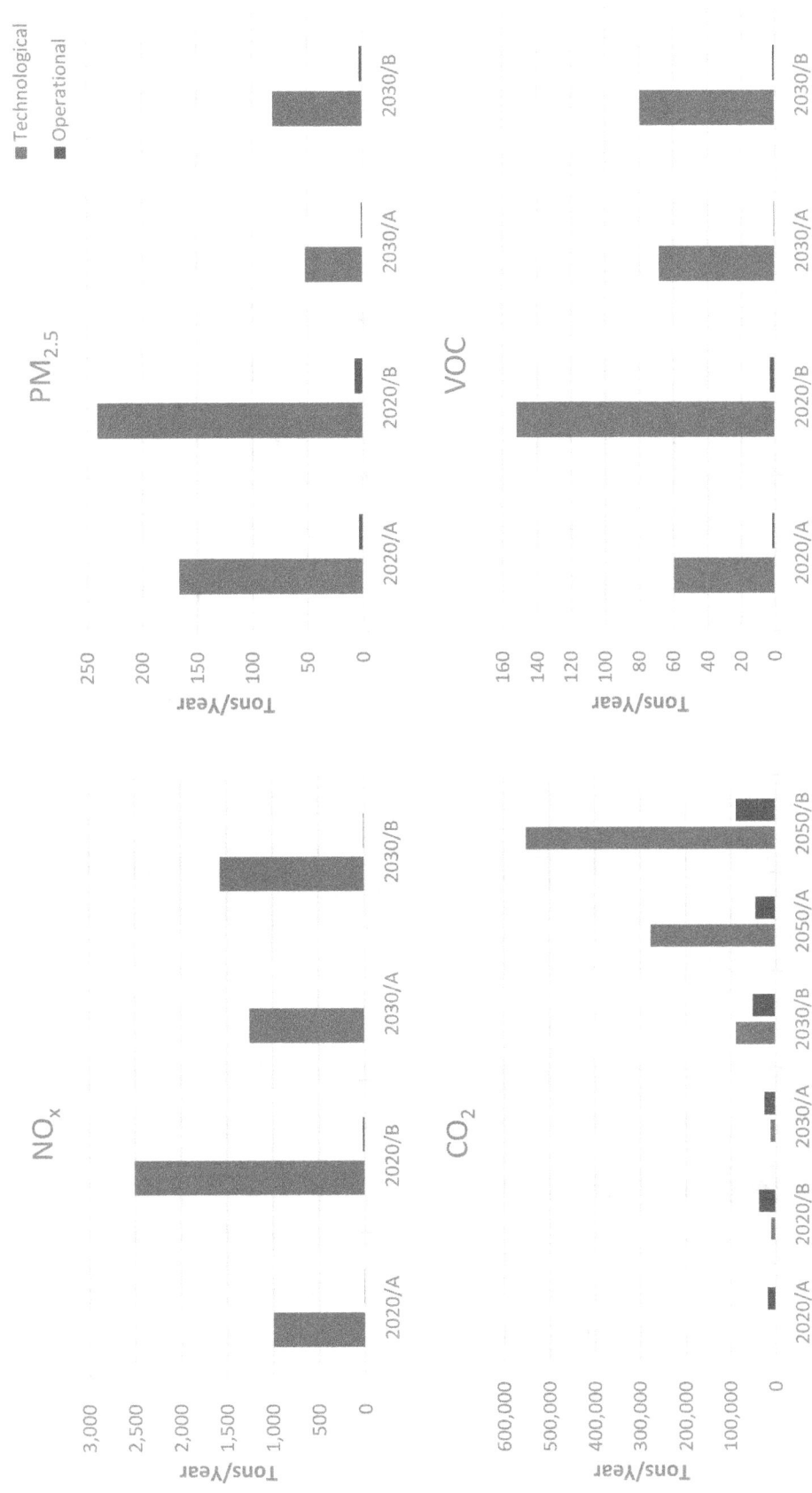

Figure C-1. Absolute Emissions Reductions from the Drayage Sector

Figure C-1. Absolute Emissions Reductions from the Drayage Sector (Continued)

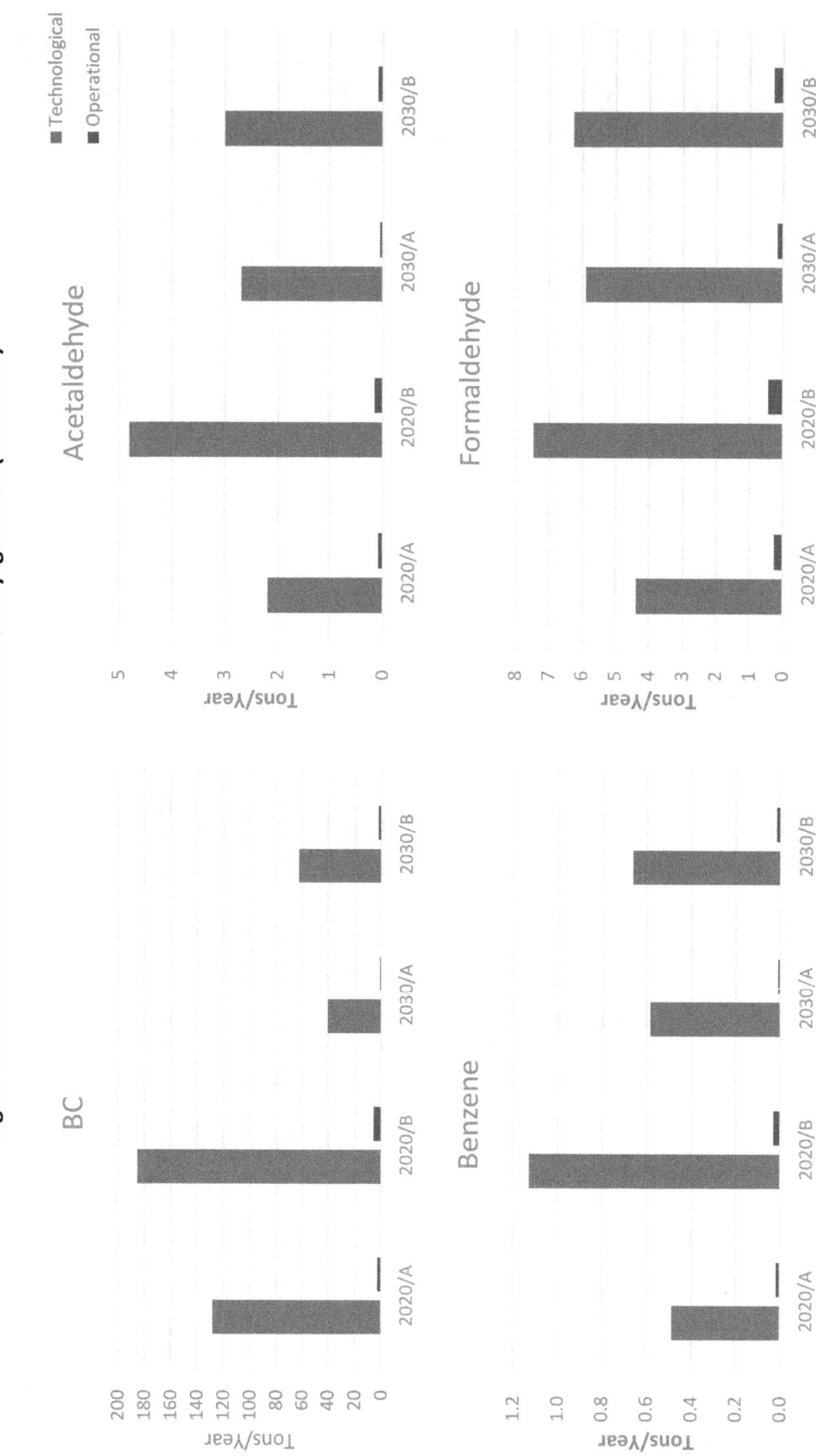

Appendix C: Analysis of Emission Reduction Scenarios

Figure C-2. Absolute Emissions Reductions from the Rail Sector

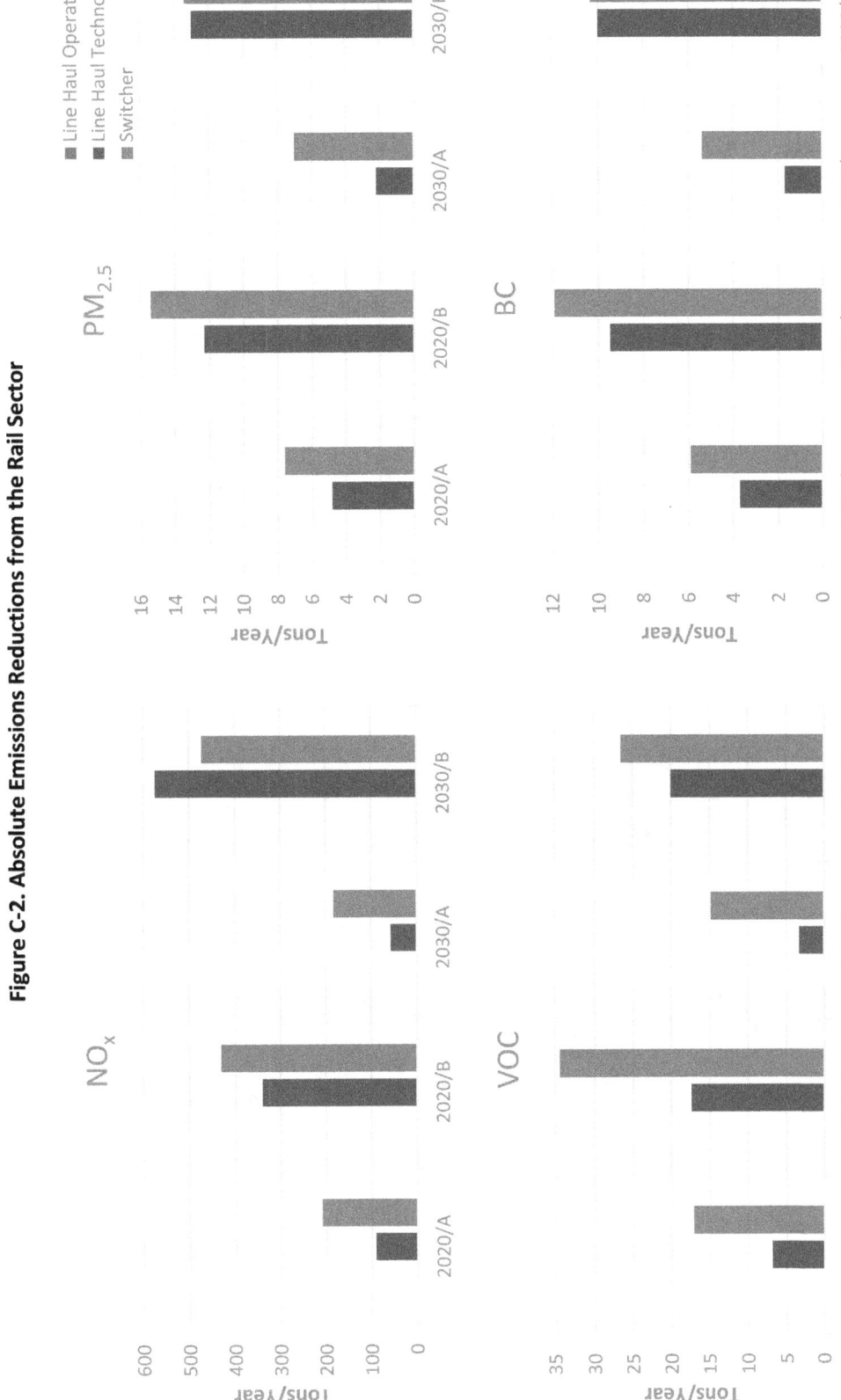

Figure C-2. Absolute Emissions Reductions from the Rail Sector (Continued)

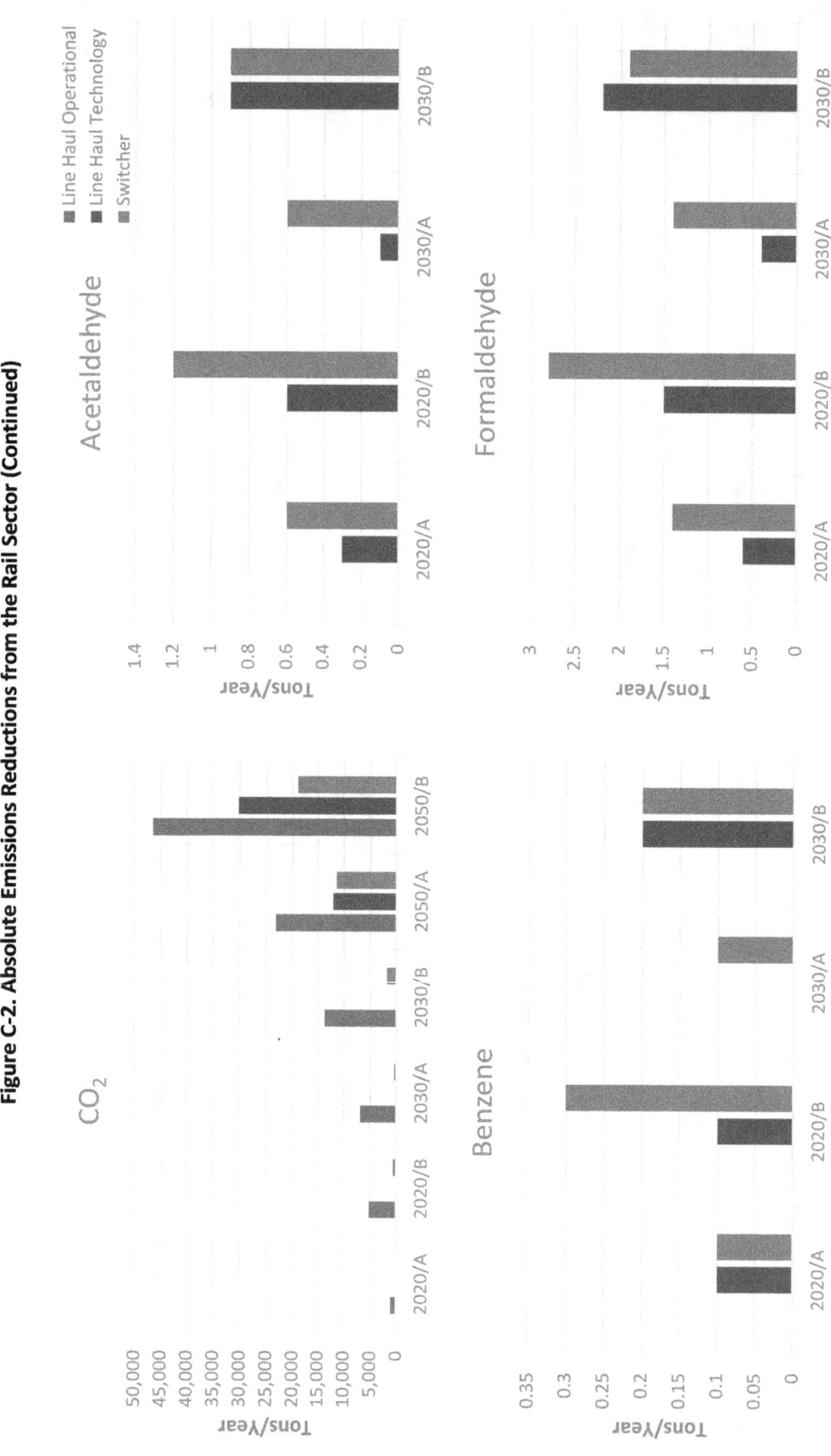

Figure C-3. Absolute Emissions Reductions from the CHE Sector

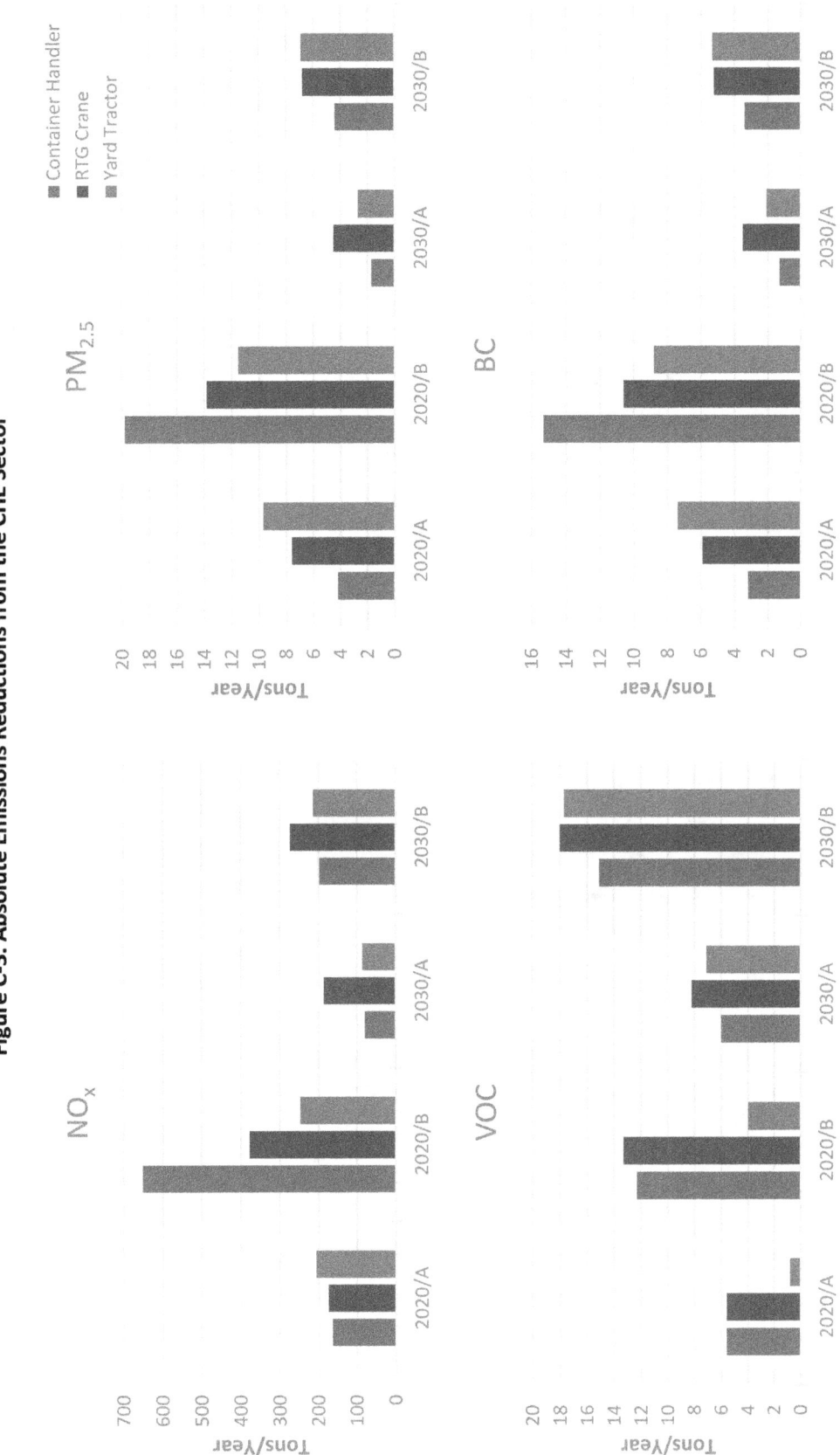

Appendix C: Analysis of Emission Reduction Scenarios

Figure C-3. Absolute Emissions Reductions from the CHE Sector (Continued)

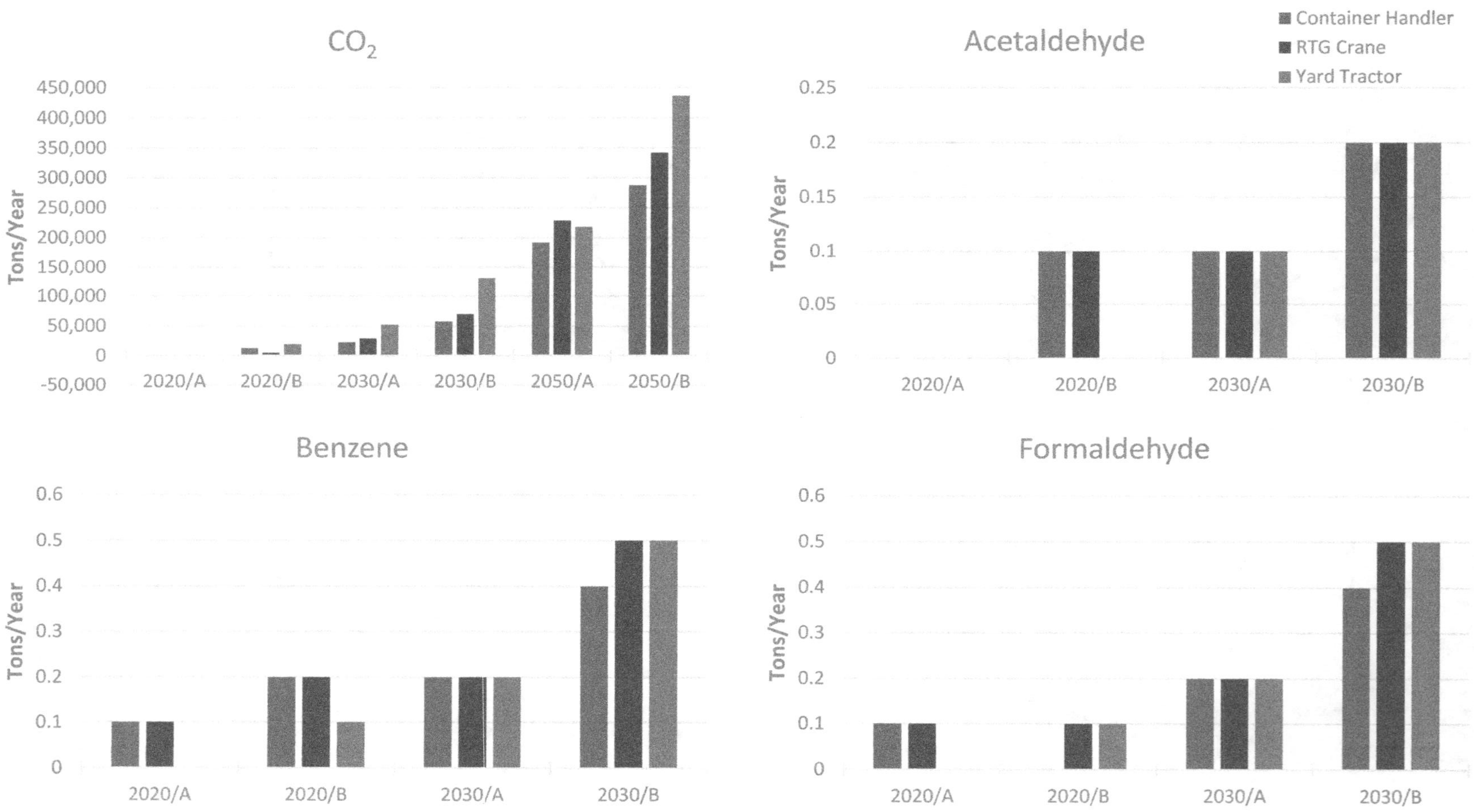

Figure C-4. Absolute Emissions Reductions from the Harbor Craft Sector

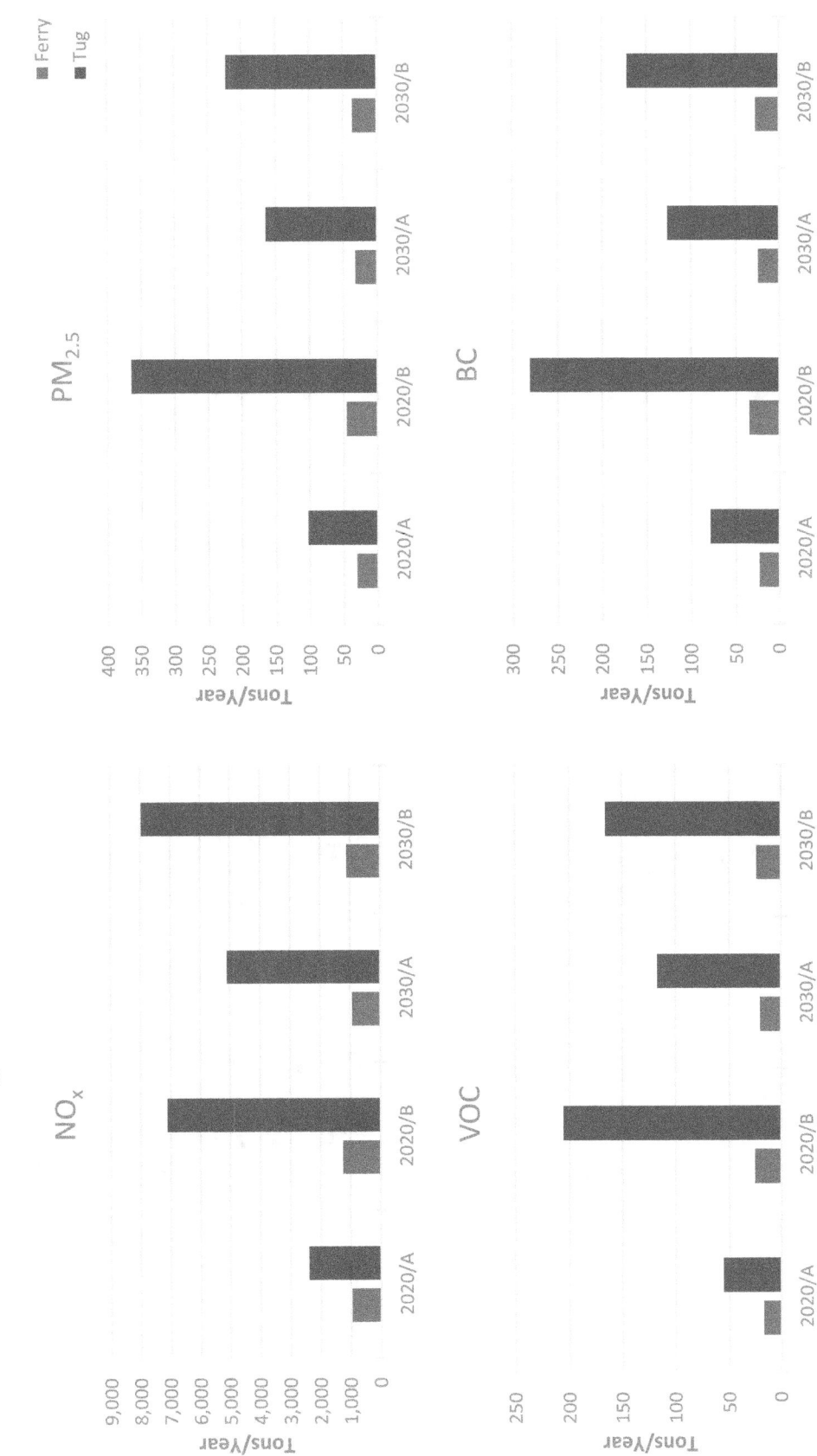

Appendix C: Analysis of Emission Reduction Scenarios

Figure C-4. Absolute Emissions Reductions from the Harbor Craft Sector (Continued)

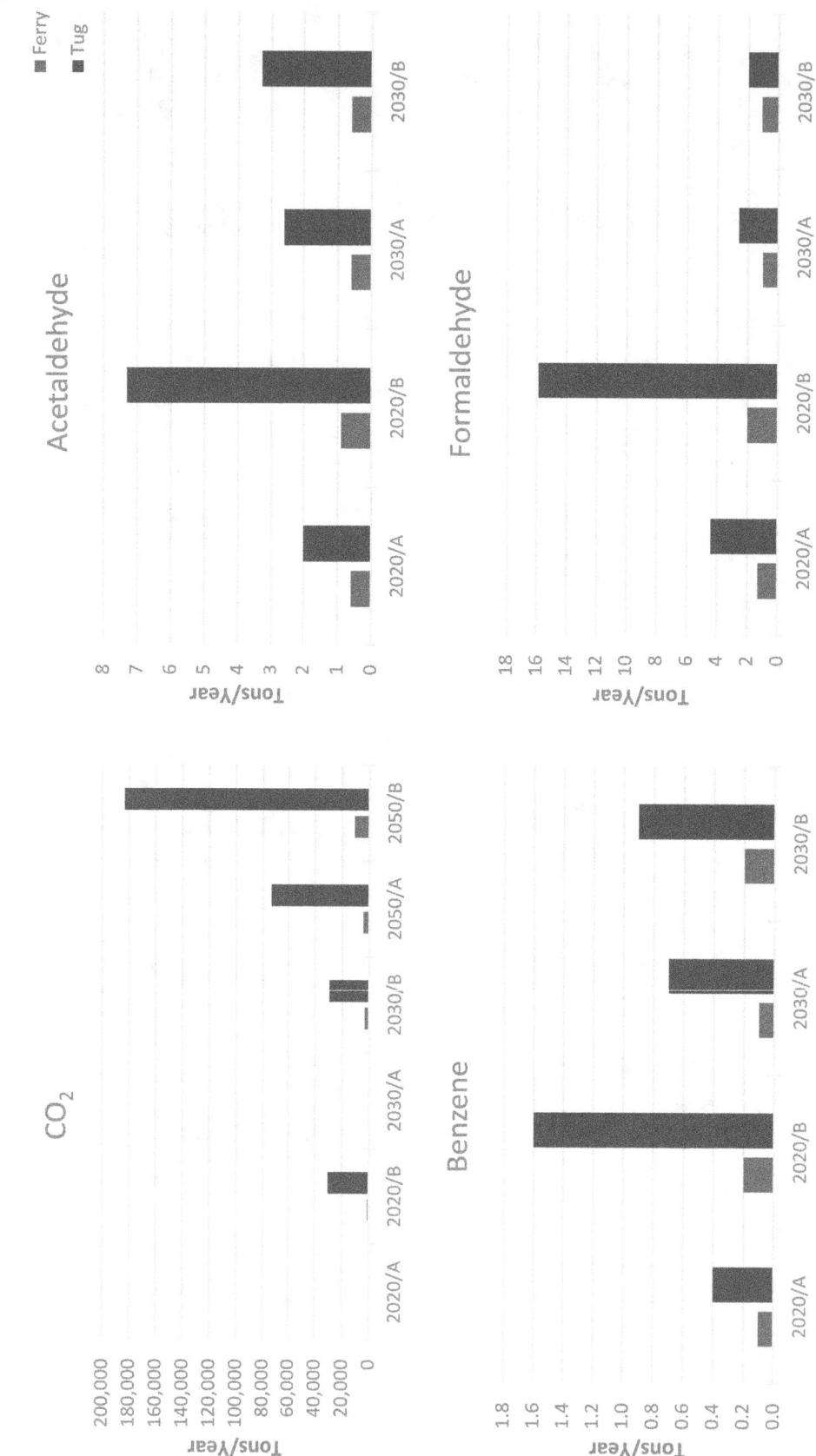

Figure C-5. Absolute Emissions Reductions from the Ocean Going Vessels Sector

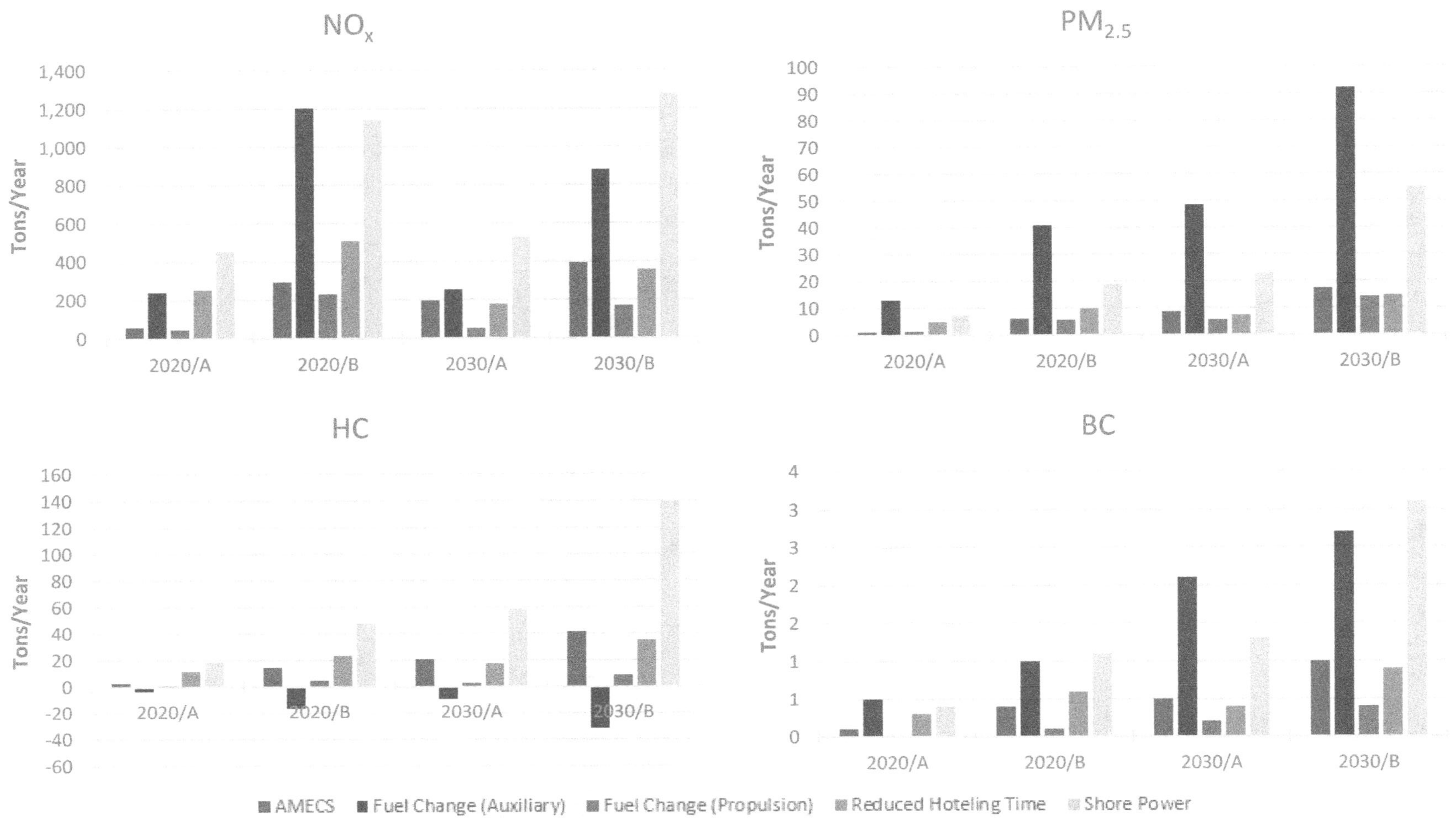

Figure C-5. Absolute Emissions Reductions from the Ocean Going Vessels Sector (Continued)

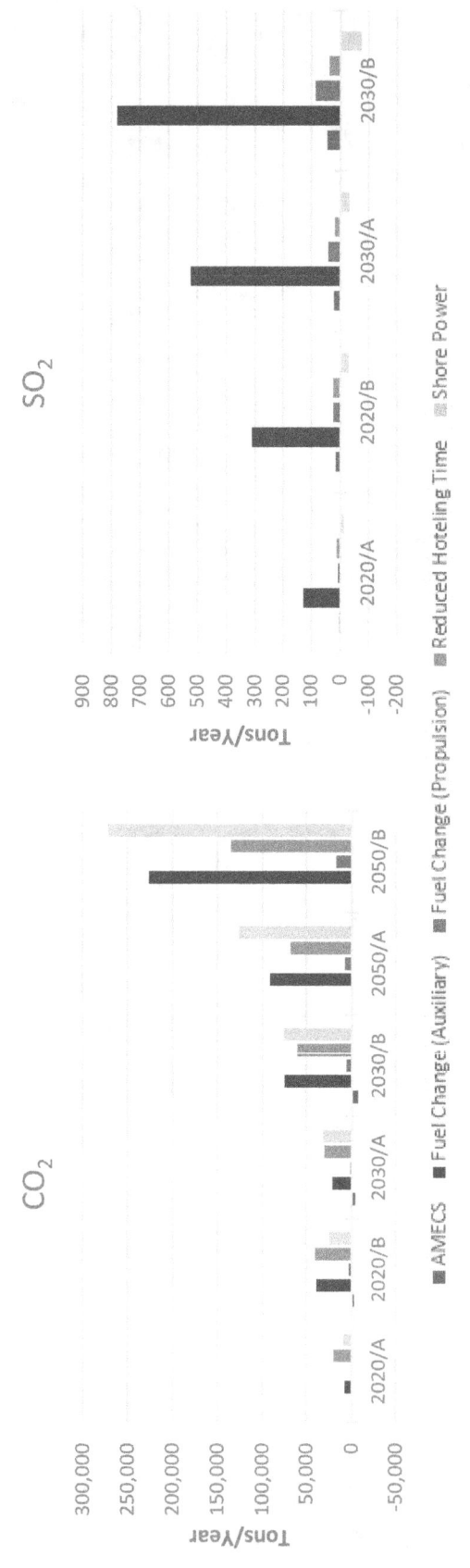

C.7.2. Relative Emission Reductions

The following figures show the percentage reductions obtained from applying each strategy as described above. Note that the reductions are relative to the applicable and relevant portion of the BAU inventories.

Figure C-6. Relative Emissions Reductions from the Drayage Sector

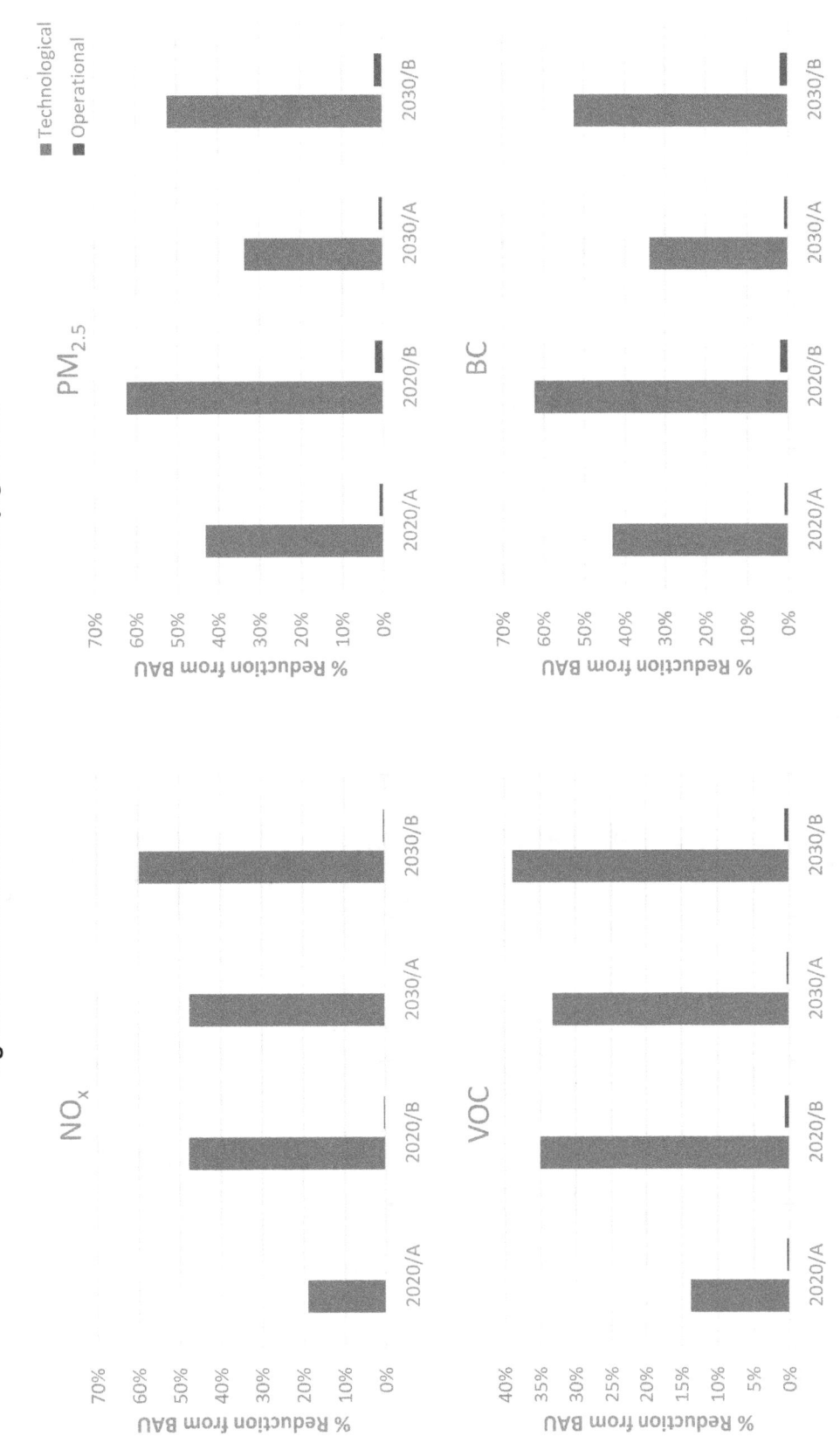

Figure C-6. Relative Emissions Reductions from the Drayage Sector (Continued)

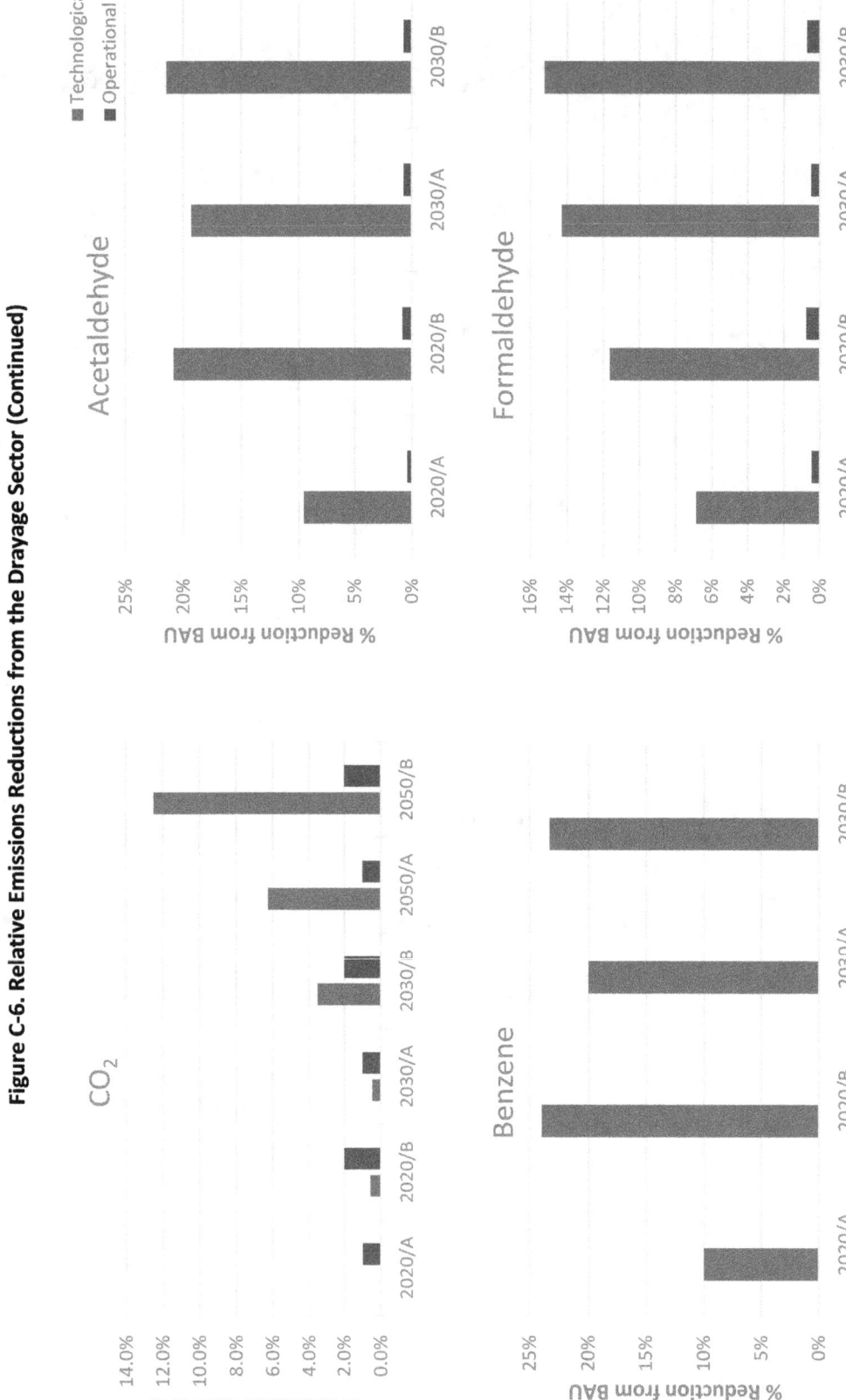

Figure C-7. Relative Emissions Reductions from the Rail Sector

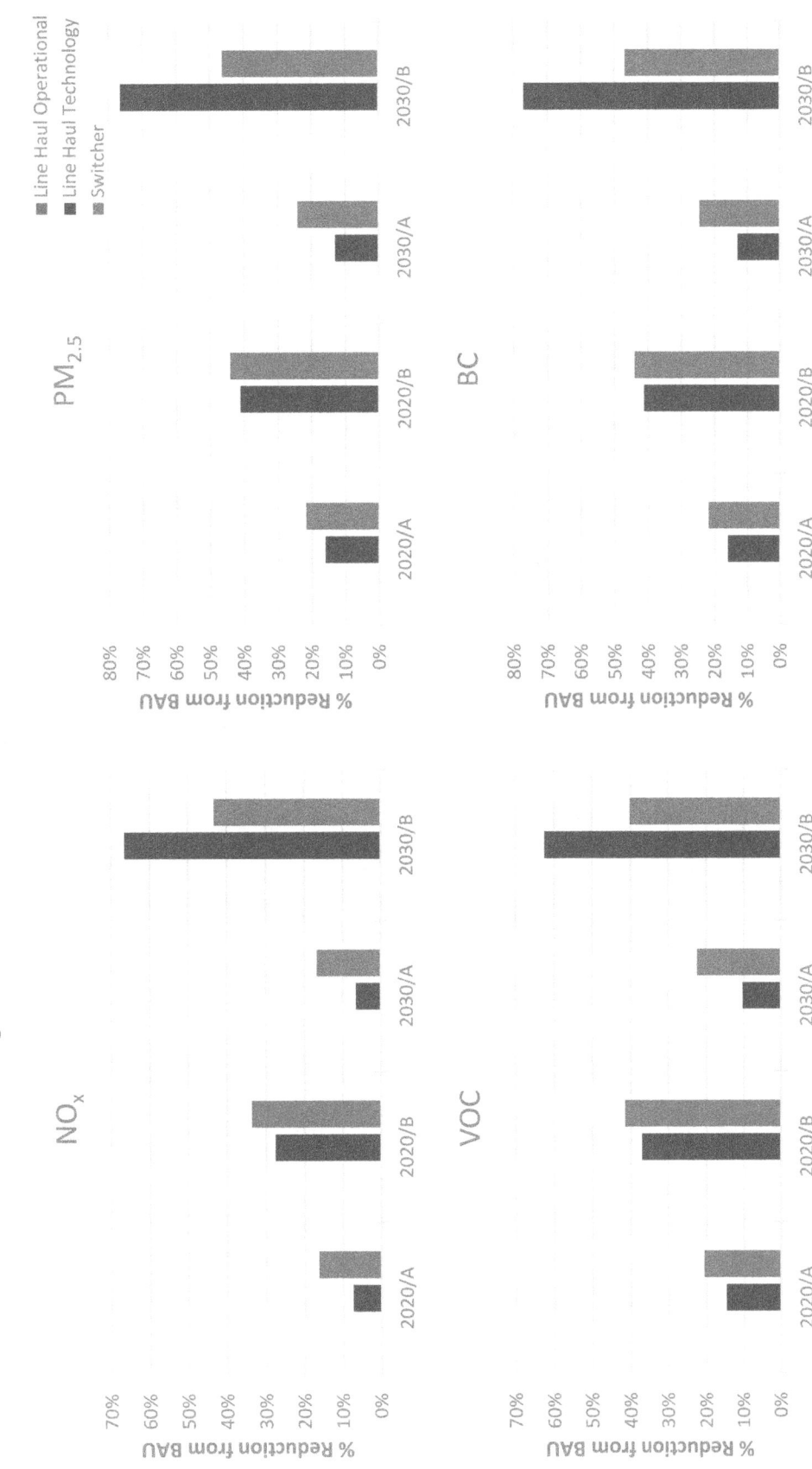

Figure C-7. Relative Emissions Reductions from the Rail Sector (Continued)

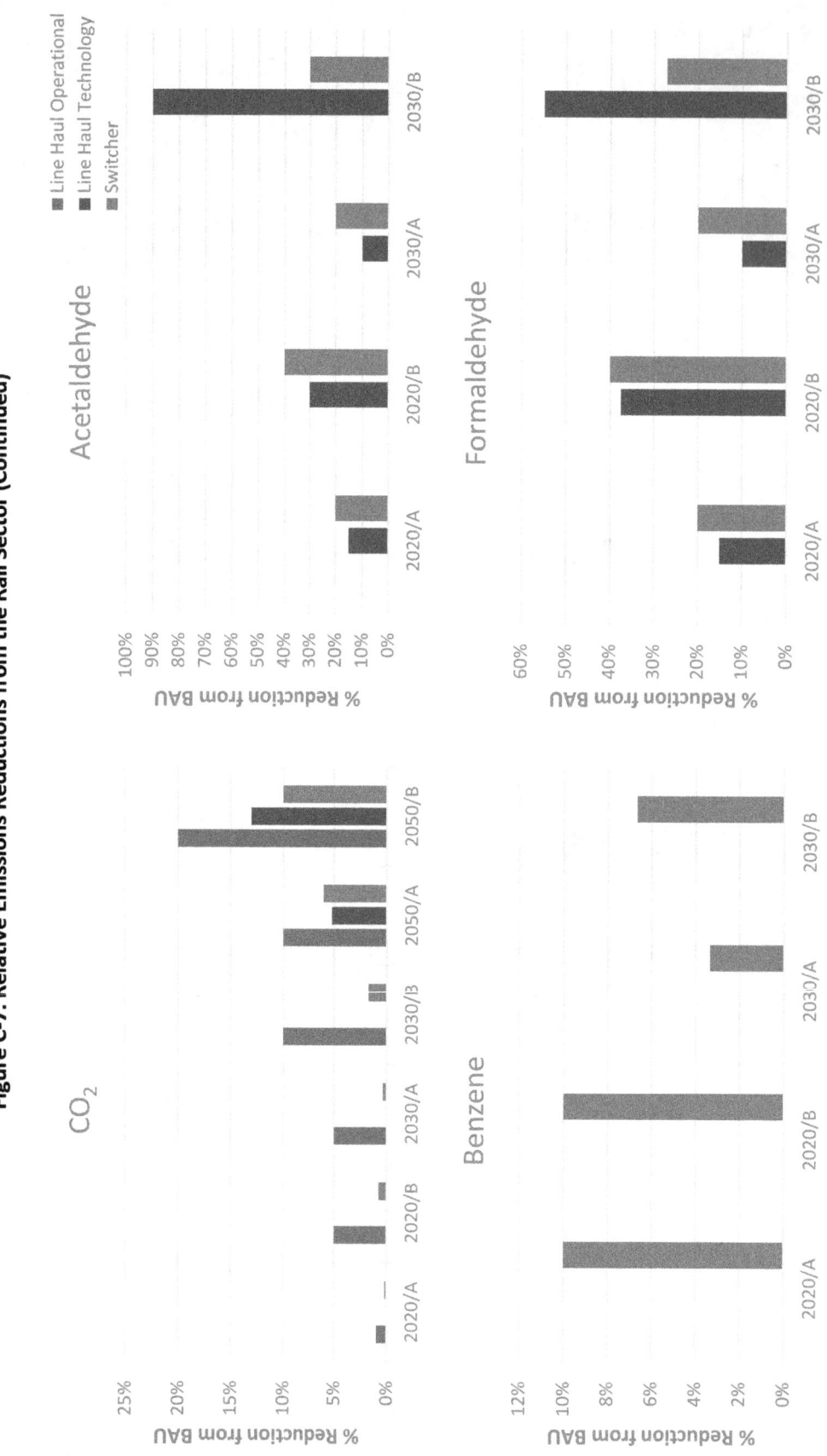

Figure C-8. Relative Emissions Reductions from the CHE Sector

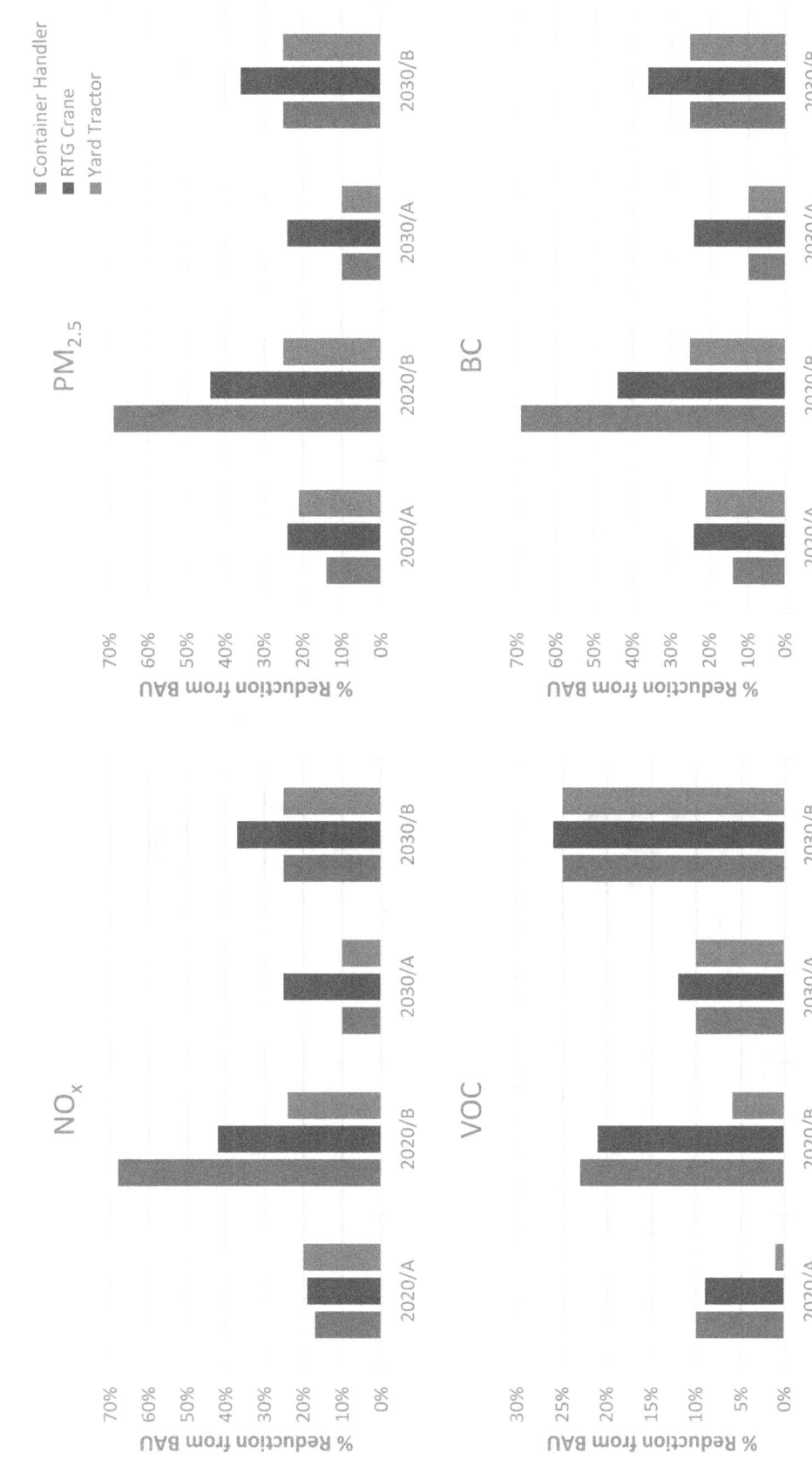

Figure C-8. Relative Emissions Reductions from the CHE Sector (Continued)

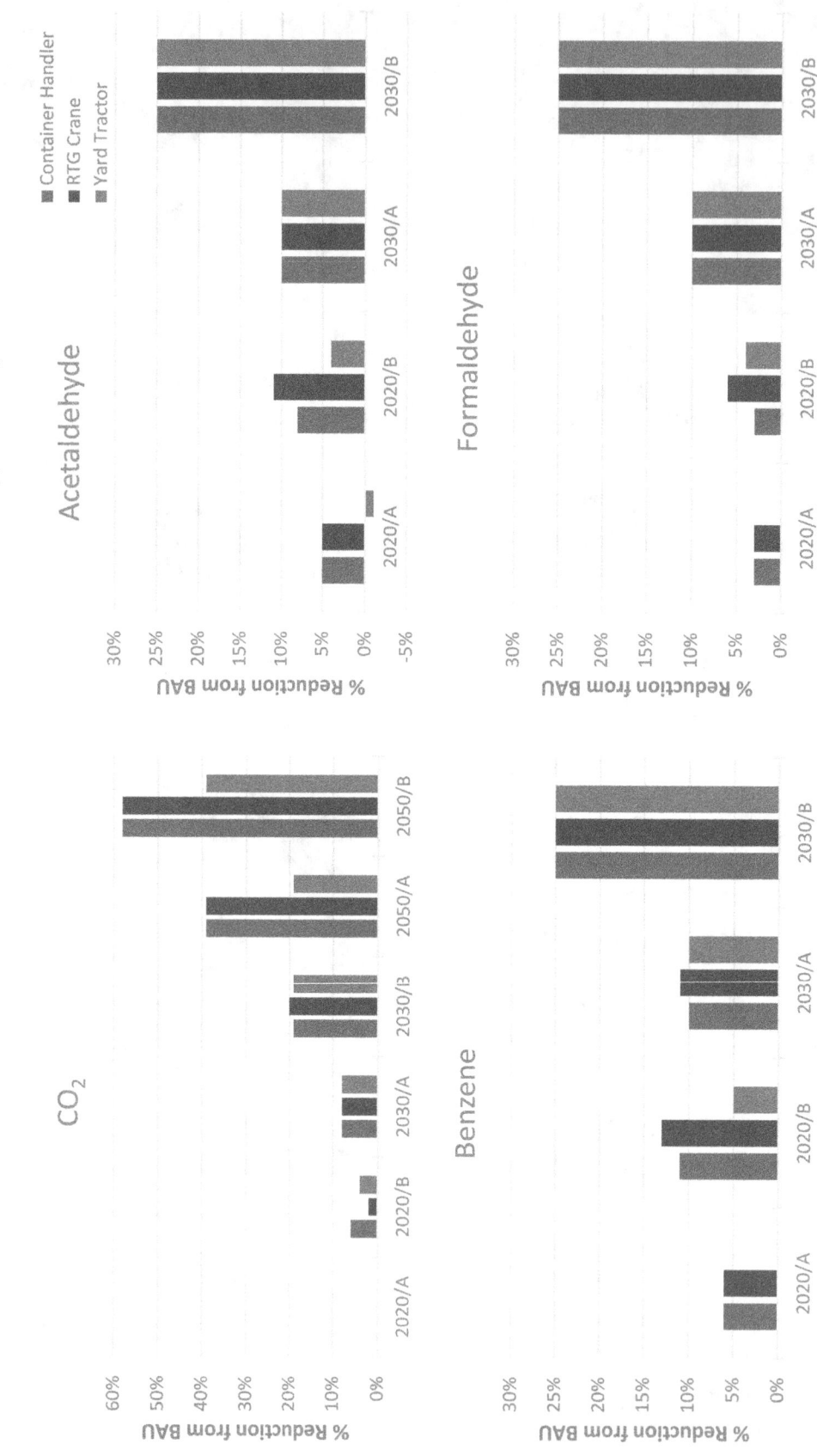

Figure C-9. Relative Emissions Reductions from the Harbor Craft Sector

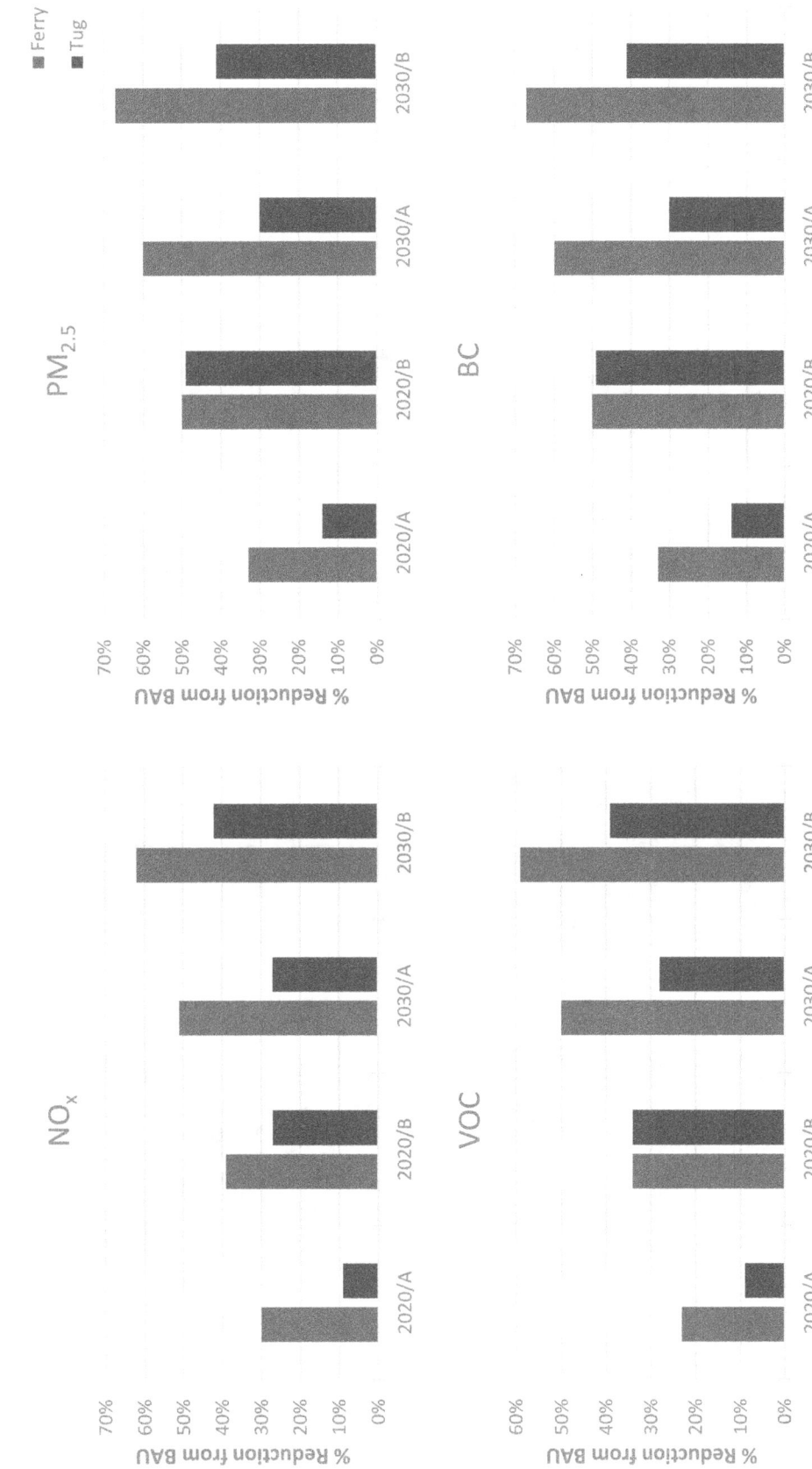

Figure C-9. Relative Emissions Reductions from the Harbor Craft Sector (Continued)

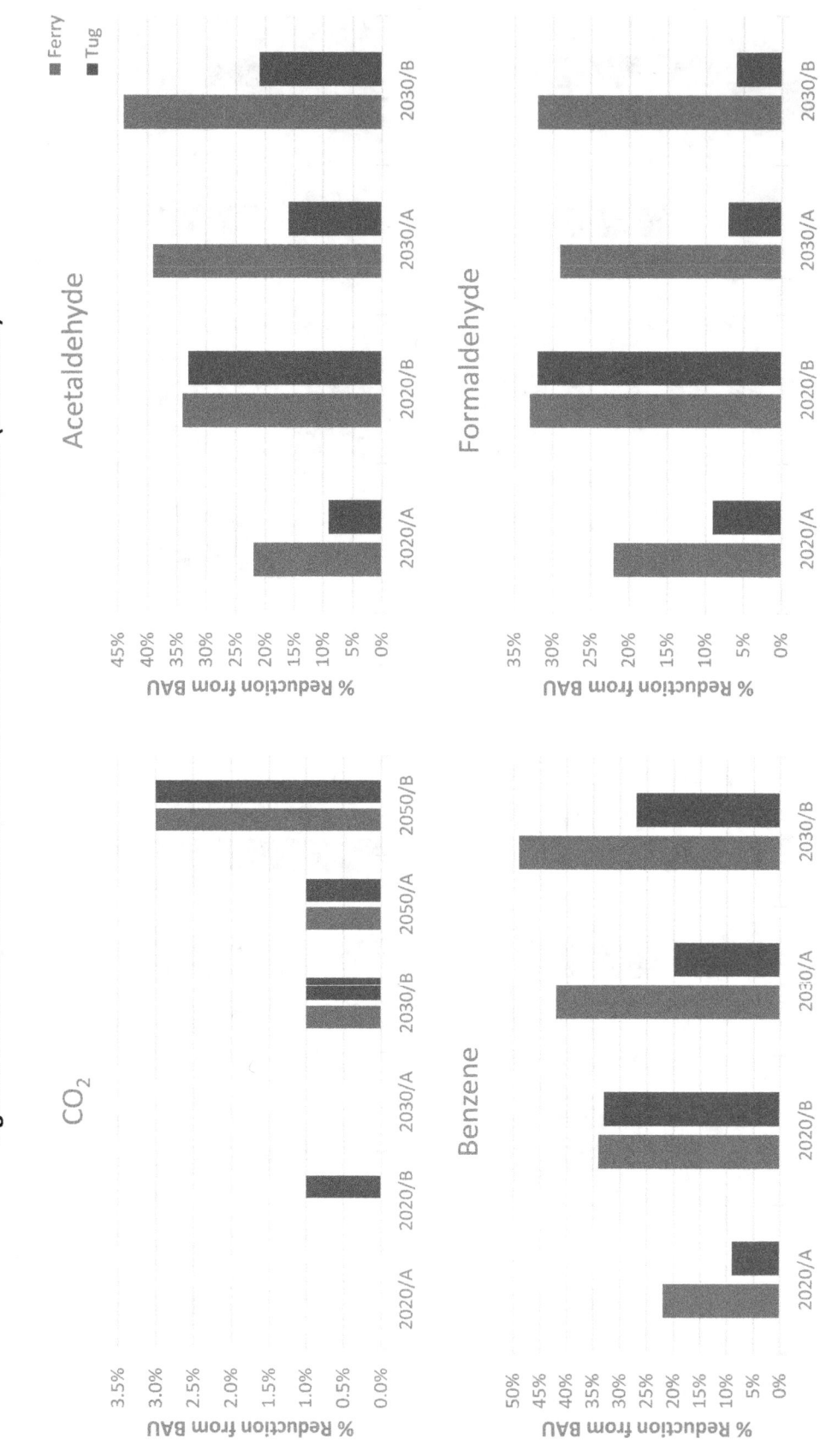

Figure C-10. Relative Emissions Reductions from the Ocean Going Vessels Sector

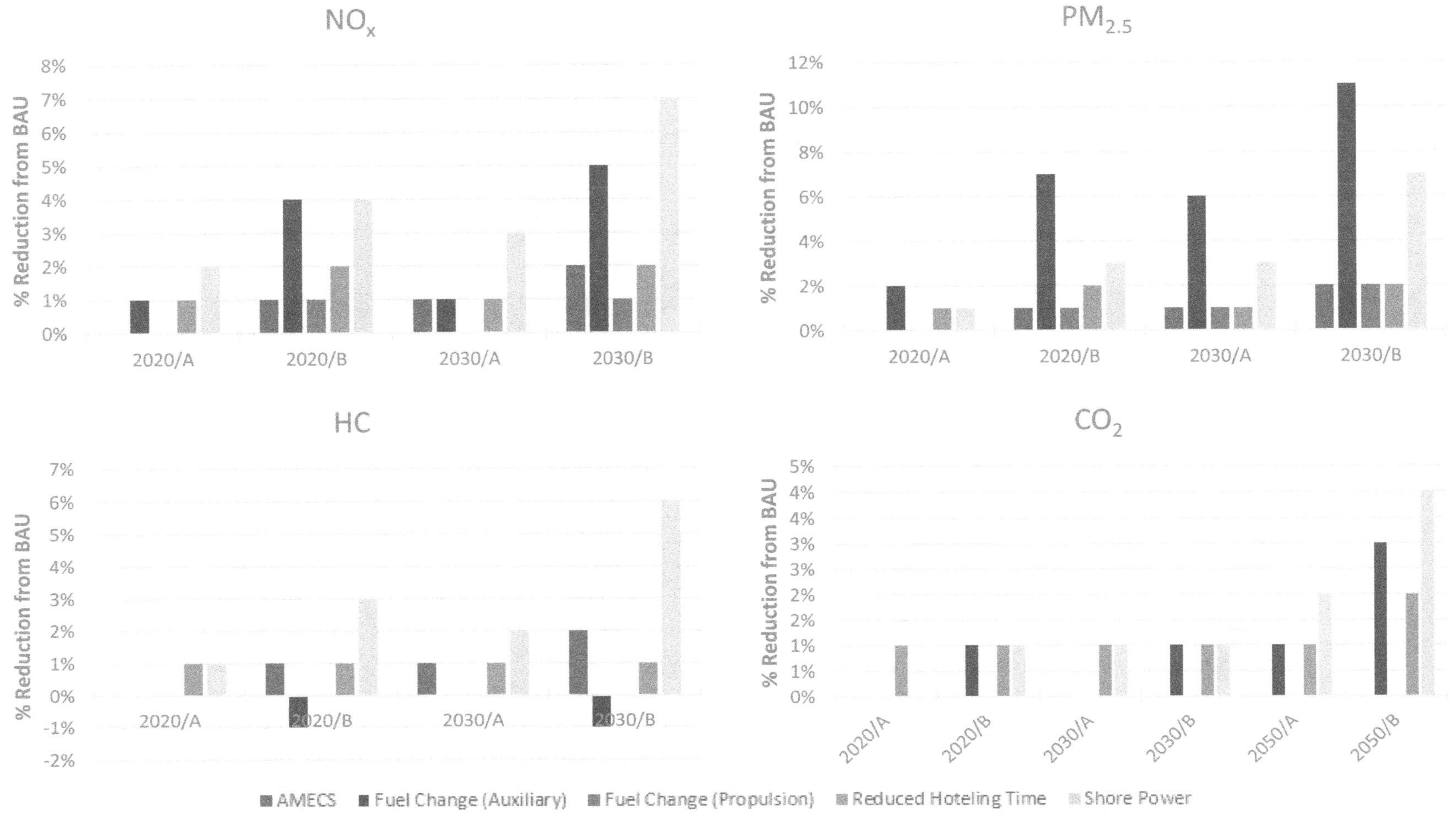

Legend: AMECS · Fuel Change (Auxiliary) · Fuel Change (Propulsion) · Reduced Hoteling Time · Shore Power

Appendix C: Analysis of Emission Reduction Scenarios

Figure C-10. Relative Emissions Reductions from the Ocean Going Vessels Sector (Continued)

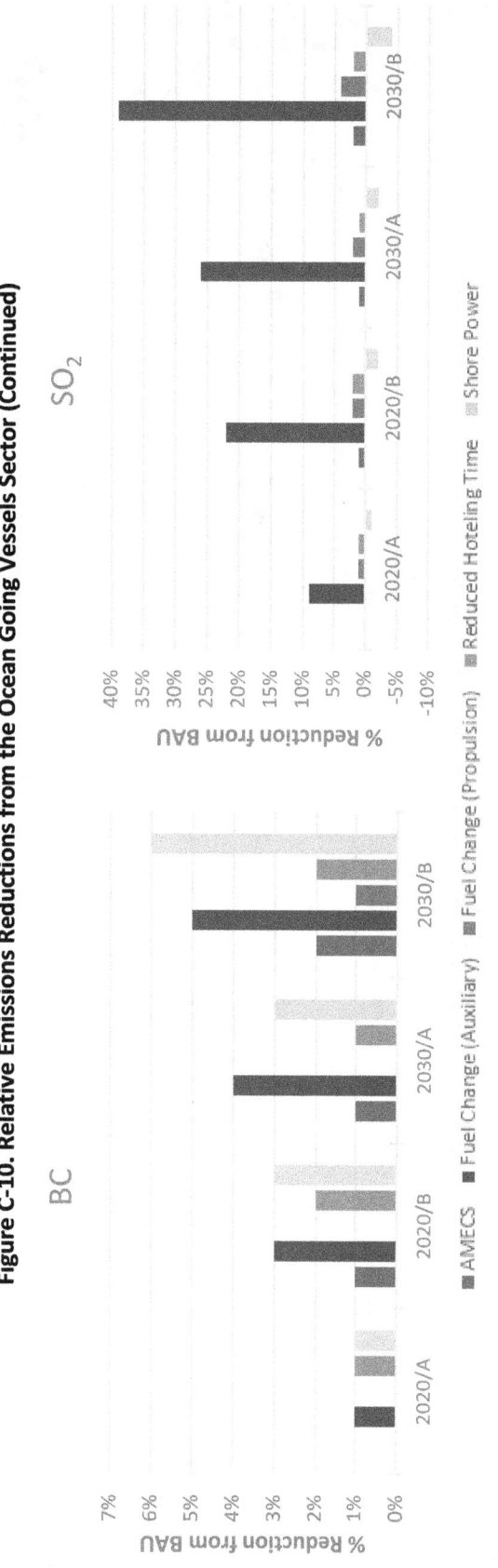

BC

SO₂

■ AMECS ■ Fuel Change (Auxiliary) ■ Fuel Change (Propulsion) ■ Reduced Hoteling Time ■ Shore Power

Appendix D.   Stratified Summary of Results

D.1. Additional Details for Stratification Analysis

The results of this assessment were stratified in a number of ways to examine which types of strategies may have more potential to reduce emissions at different kinds of ports. This analysis was performed separately for the OGV and non-OGV sectors due to the nature of the various strategies applied in each sector. This appendix provides more details on some aspects of this analysis as well as additional charts of stratification results.

The ports were classified as "container" if their cargo throughput was greater than 100,000 twenty-foot equivalent units (TEUs). Container ports were further classified as "small" if their cargo throughput was less than 1 million TEUs and "large" if it was more. Additionally, ports were considered "bulk" if their non-container throughput was greater than 20,000 tons per year (tpy); the cutoff between large and small bulk ports was 50,000 tpy. Finally, ports were classified as "passenger" based on engineering judgement. Large passenger ports were ports with more than 750,000 annual passengers. The TEU and tonnage data by port came from U.S. Army Corps of Engineers' Waterborne Commerce Statistics Center.

Each classification was made independently of the others, so that each port might fall into any number of categories and may have different size distinctions. For example, a port could be labeled as both a small passenger port and a large container port. However, it is important to note that these classifications and distinctions are not official determinations, but are simply used in the stratification analysis to differentiate generally between the different kinds of ports included in this assessment. The distinctions "large" and "small" only serve to compare between ports in this assessment and do not facilitate other comparisons. The cutoff points between the two distinctions were chosen such that the large and small ports within a classification contained a roughly equal number of ports.

A listing of ports that fall under each type and size category may be found in Table D-1. This analysis aggregated the emissions across these 19 port areas to examine the potential impacts of emission reduction strategy scenarios at the national scale; this assessment (including the stratification analysis) does not provide specific data for local decision-making at individual ports or specific neighborhoods.

Table D-1. Port Classification by Type and Size for Stratification Analysis Only

Type	Size	Port
Container	Large	Port of New York and New Jersey
		Port of South Louisiana
		Port of Savannah
		Port of Seattle
		Port of Hampton Roads (Norfolk)
		Port of Houston
		Port of Charleston
	Small	Port of San Juan, PR
		Port of Miami
		Port of Baltimore
		Port of New Orleans
		Port of Philadelphia
		Port of Portland, OR
		Port of Mobile
Bulk	Large	Port of South Louisiana
		Port of Houston
		Port of New York and New Jersey
		Port of New Orleans
		Port of Corpus Christi
		Port of Baton Rouge
		Port of Mobile
	Small	Port of Baltimore
		Port of Hampton Roads (Norfolk)
		Port of Savannah
		Port Tampa Bay
		Port of Philadelphia
		Port Arthur
		Port of Portland, OR
Passenger	Large	Port of Miami
		Port of New Orleans
		Port Tampa Bay
		Port of Seattle
	Small	Port of New York and New Jersey
		Port of Baltimore
		Port of San Juan, PR

D.2. Additional Charts for Stratification Analysis of OGV Sector

D.2.1. Entire OGV Sector

Figure D-1. Comparing NOx Relative Reduction Potential of the OGV Sector

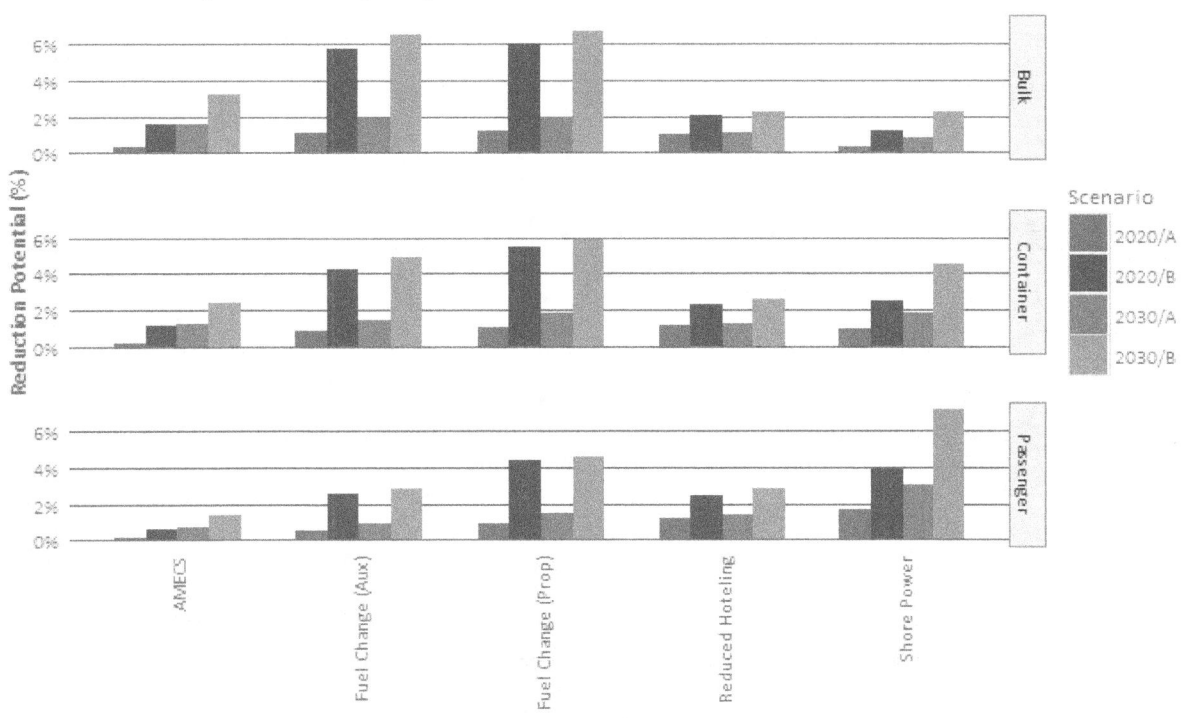

Figure D-2. Comparing PM2.5 Relative Reduction Potential of the OGV Sector

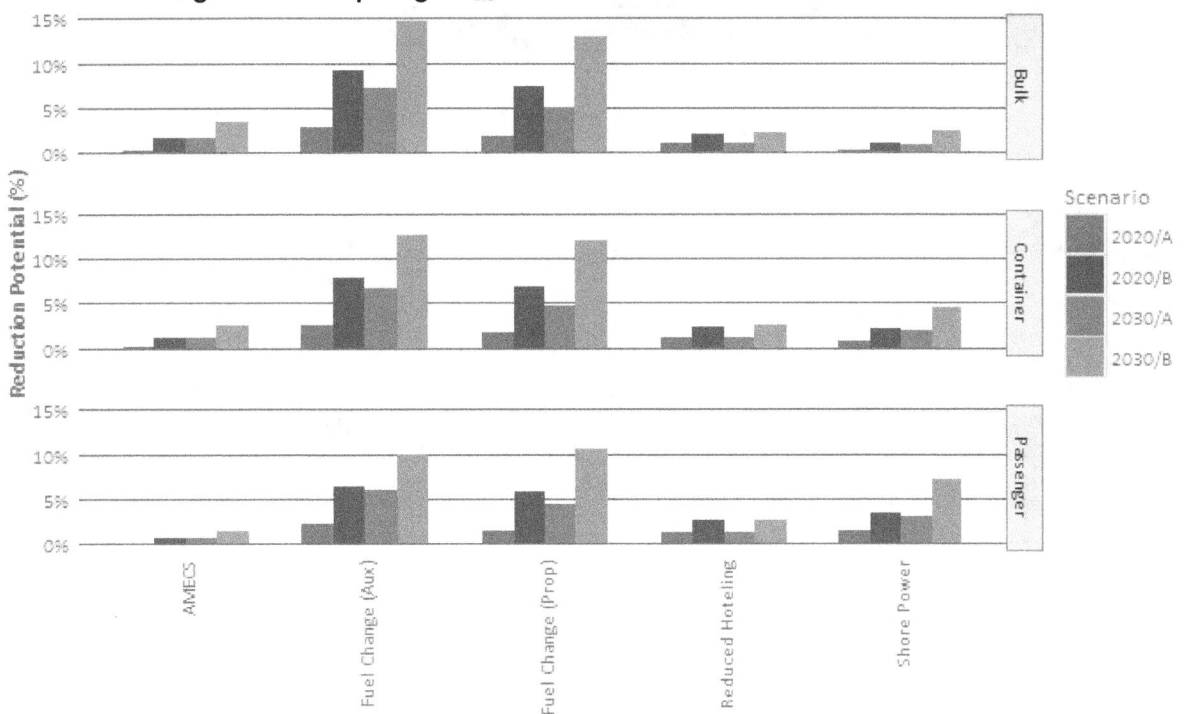

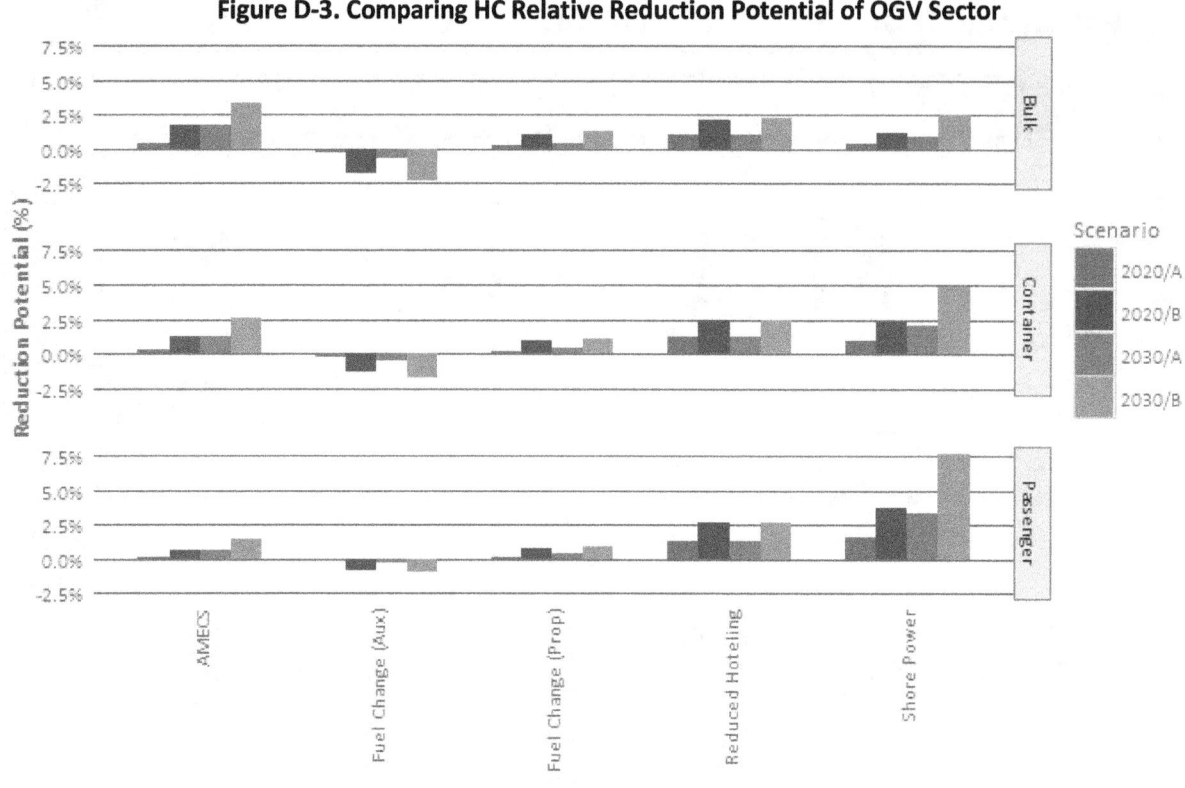

Figure D-3. Comparing HC Relative Reduction Potential of OGV Sector

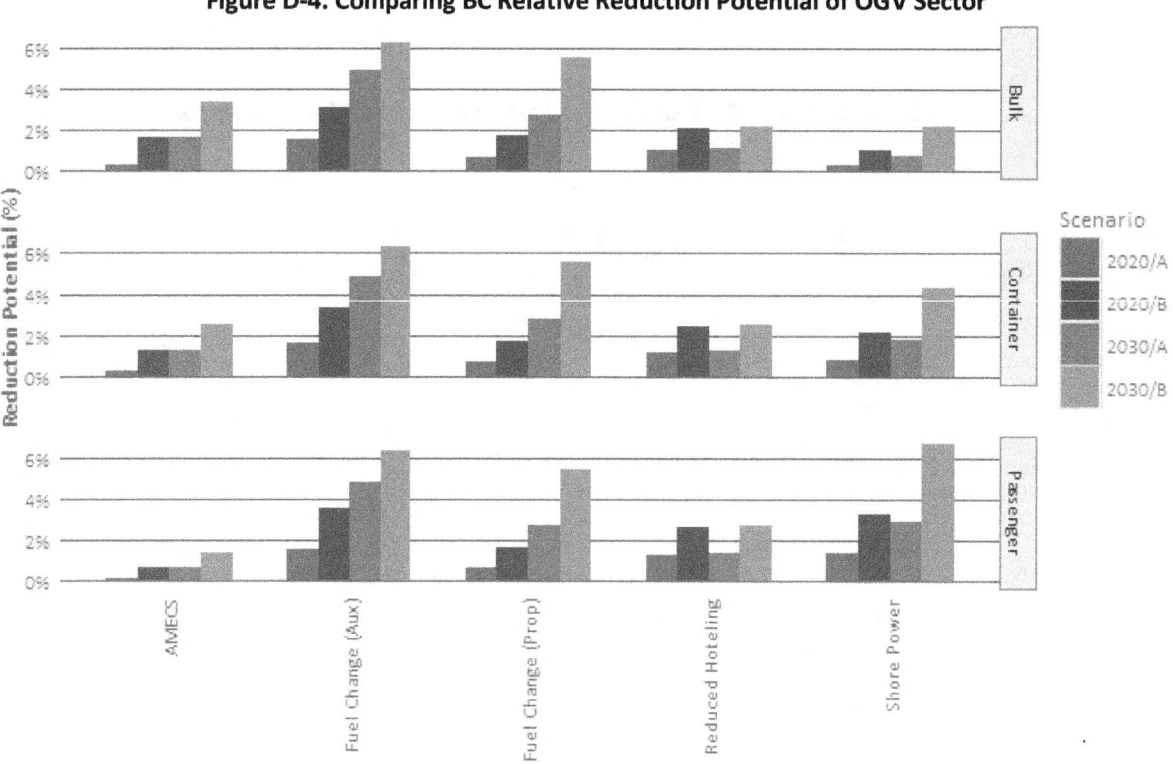

Figure D-4. Comparing BC Relative Reduction Potential of OGV Sector

Figure D-5. Comparing SO₂ Relative Reduction Potential of OGV Sector

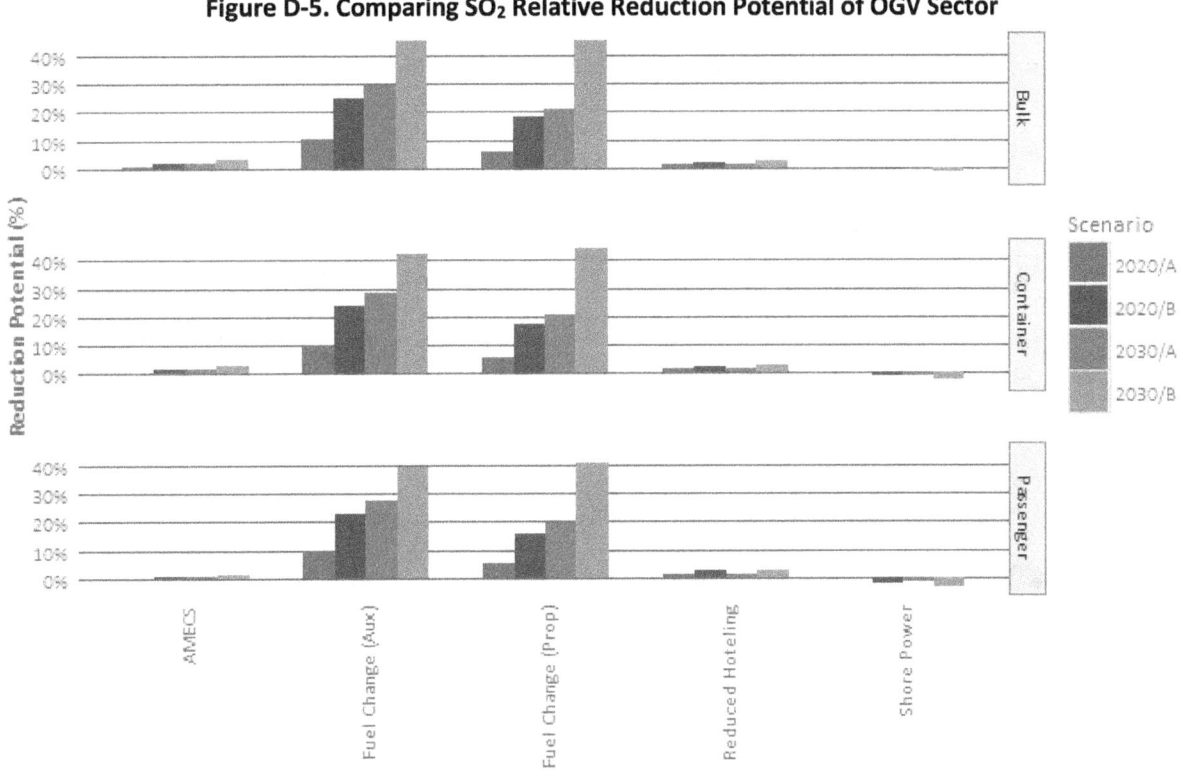

Figure D-6. Comparing CO₂ Relative Reduction Potential of OGV Sector

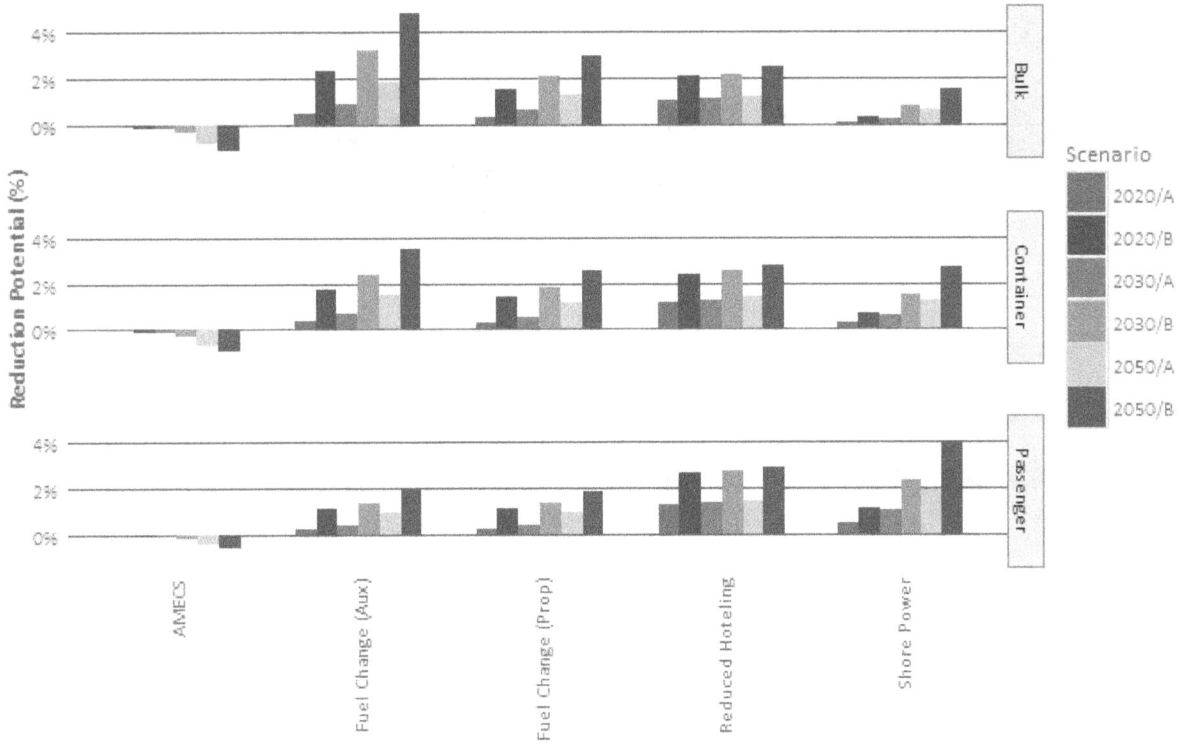

D.2.2. Container Ports

Figure D-7. NOx Relative Reduction Potential of the OGV Sector for Container Ports

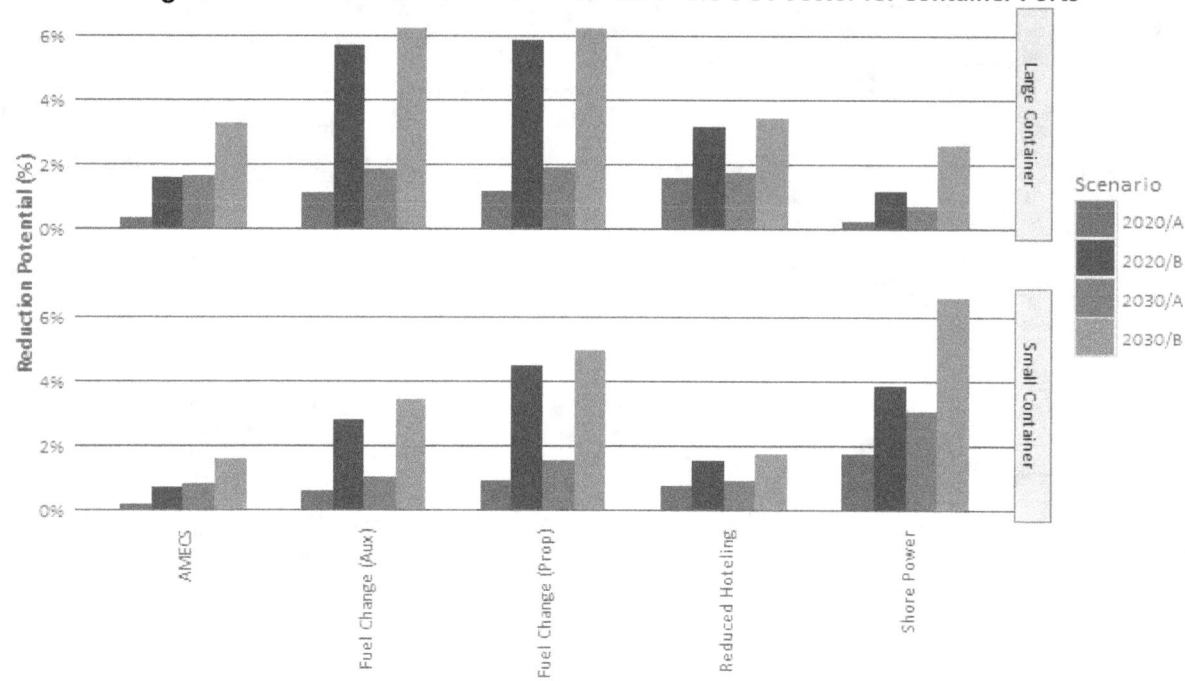

Figure D-8. PM₂.₅ Relative Reduction Potential of the OGV Sector for Container Ports

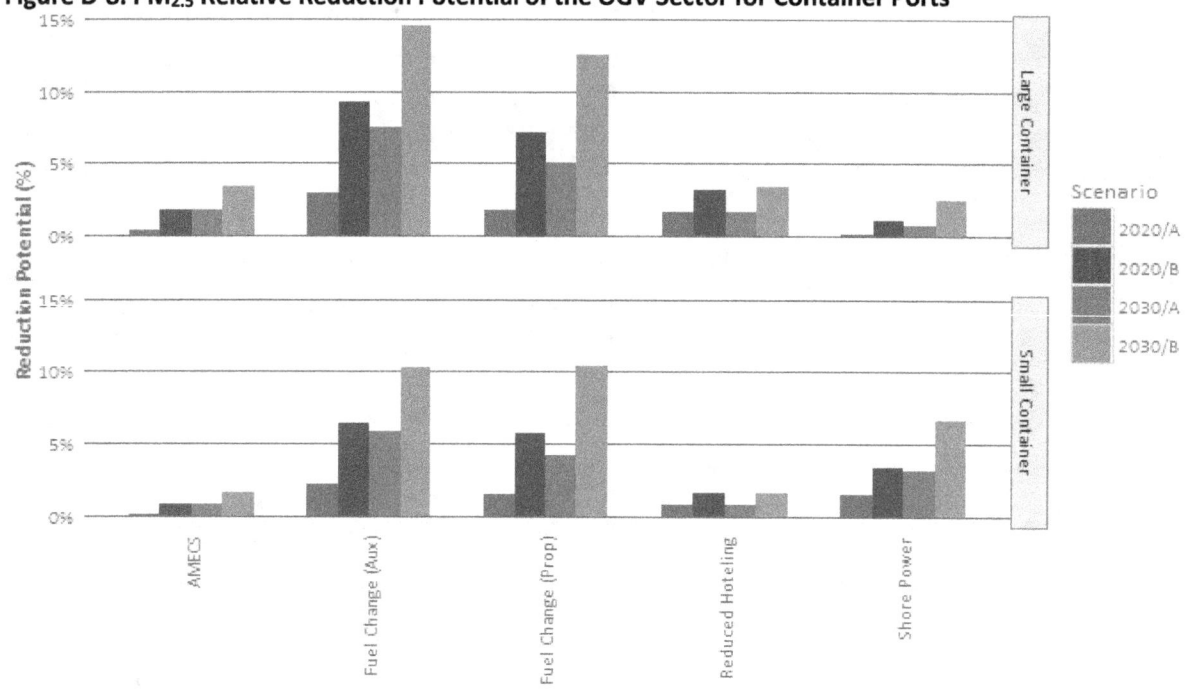

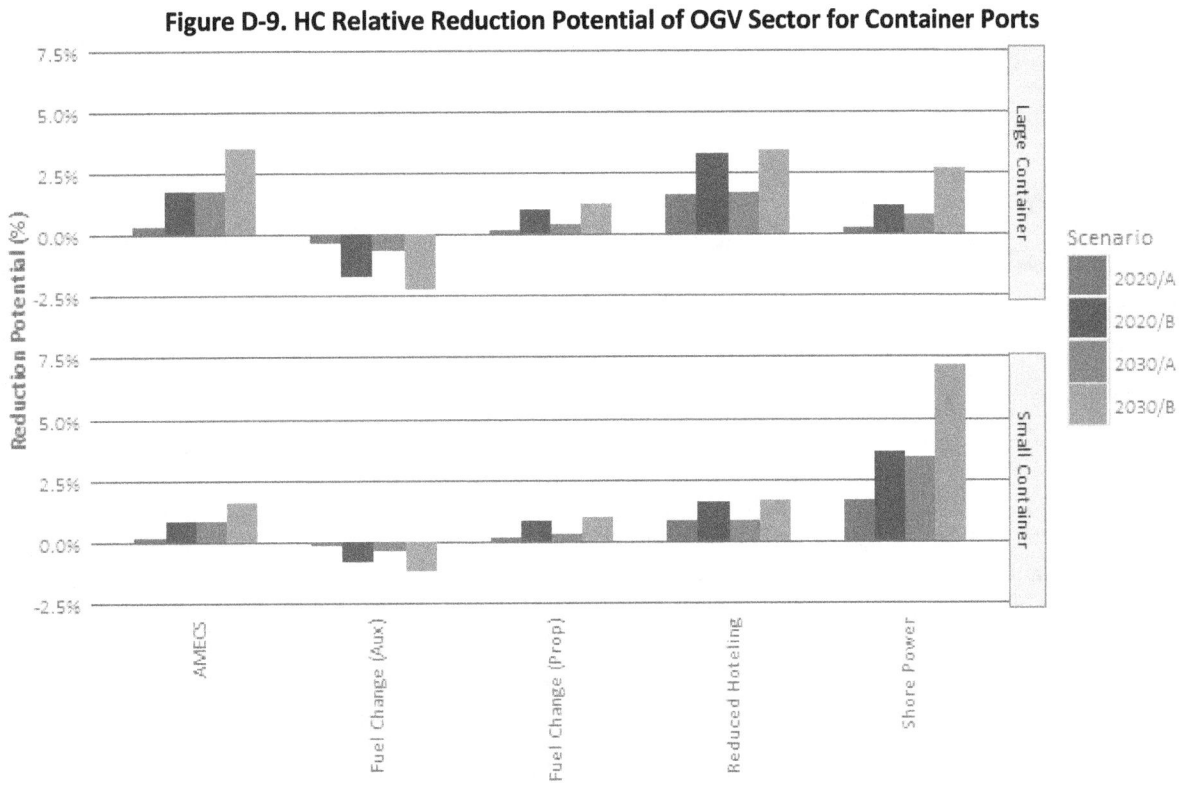

Figure D-9. HC Relative Reduction Potential of OGV Sector for Container Ports

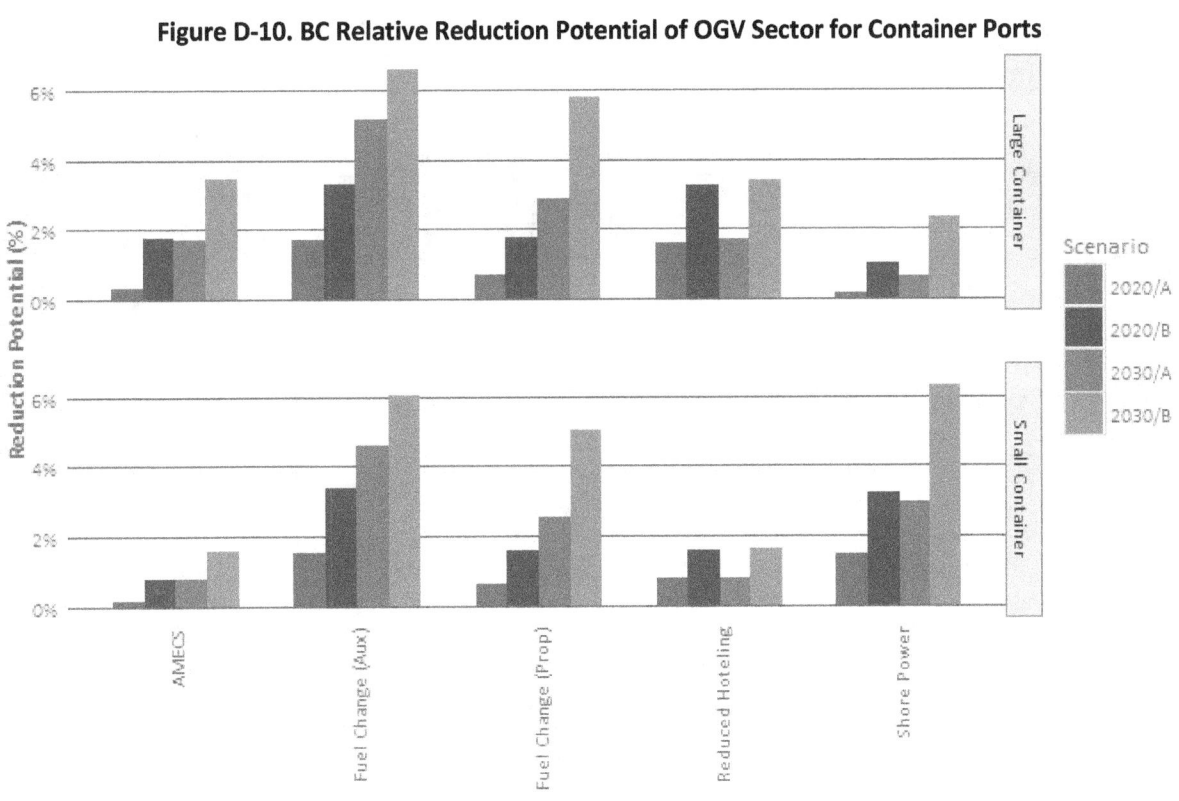

Figure D-10. BC Relative Reduction Potential of OGV Sector for Container Ports

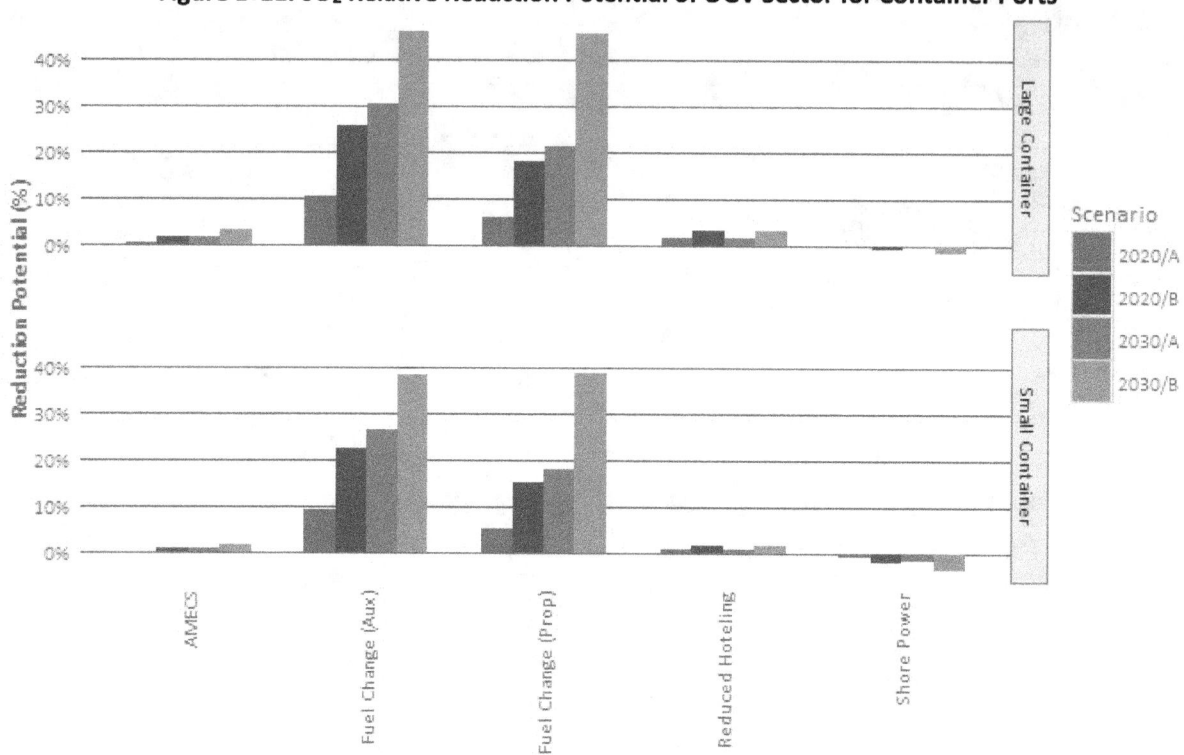

Figure D-11. SO₂ Relative Reduction Potential of OGV Sector for Container Ports

Figure D-12. CO₂ Relative Reduction Potential of OGV Sector for Container Ports

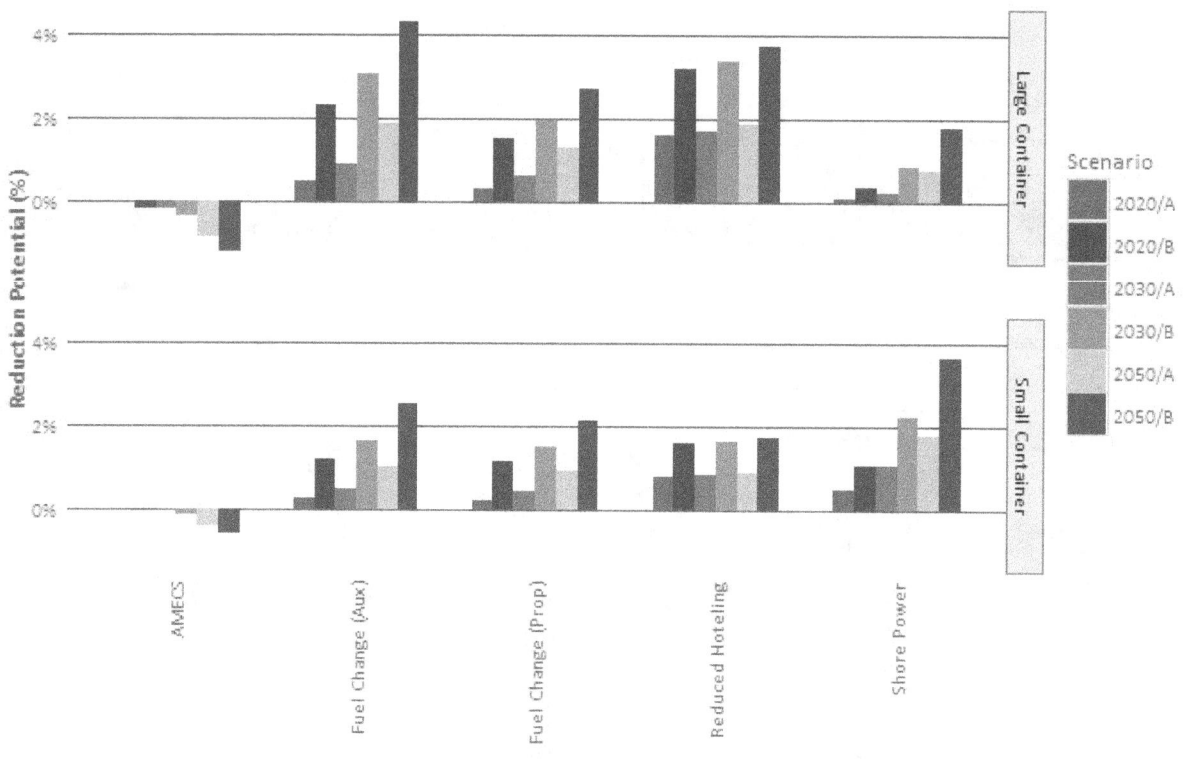

D.2.3. Bulk Ports

Figure D-13. NOx Relative Reduction Potential of OGV Sector for Bulk Ports

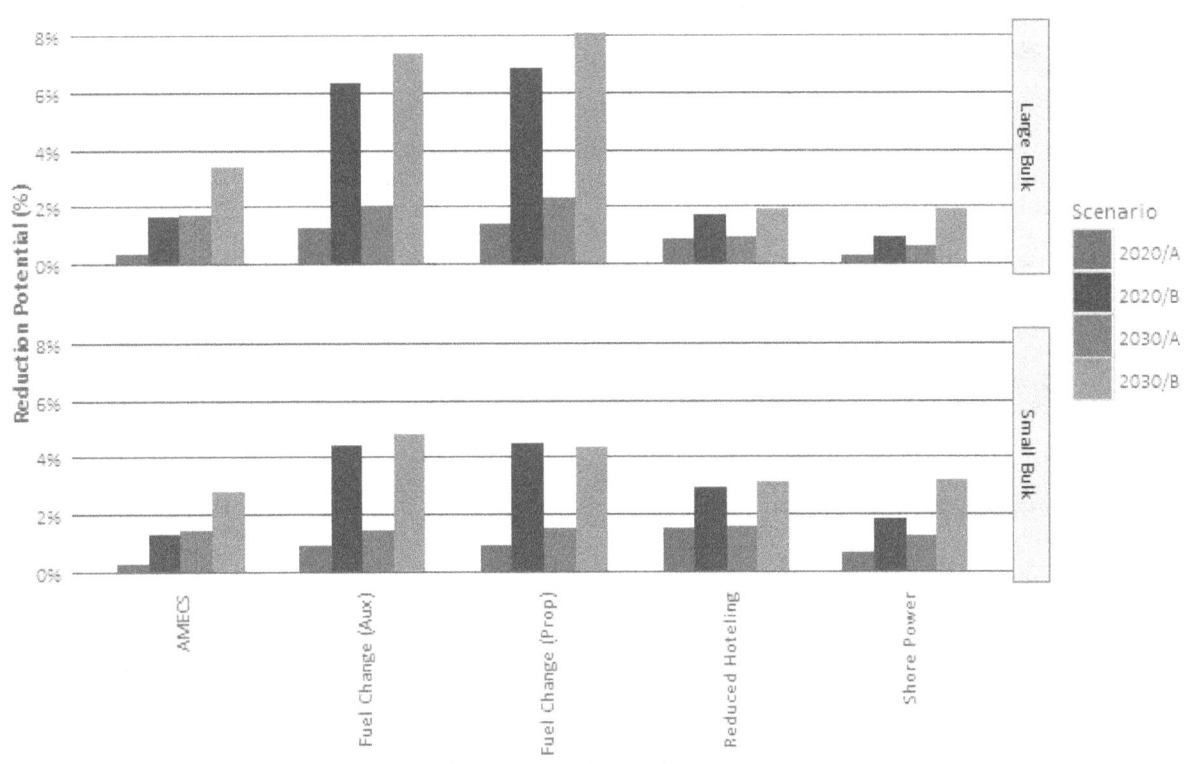

Figure D-14. PM Relative Reduction Potential of OGV Sector for Bulk Ports

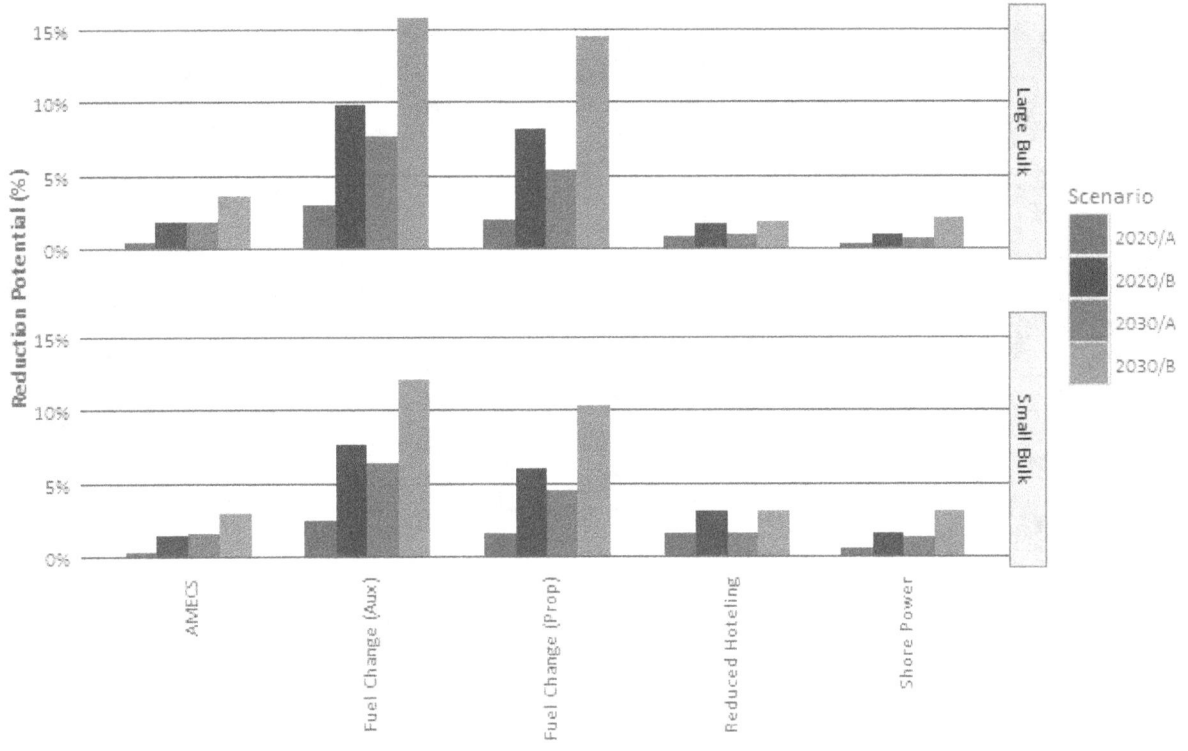

Figure D-15. HC Relative Reduction Potential of OGV Sector for Bulk Ports

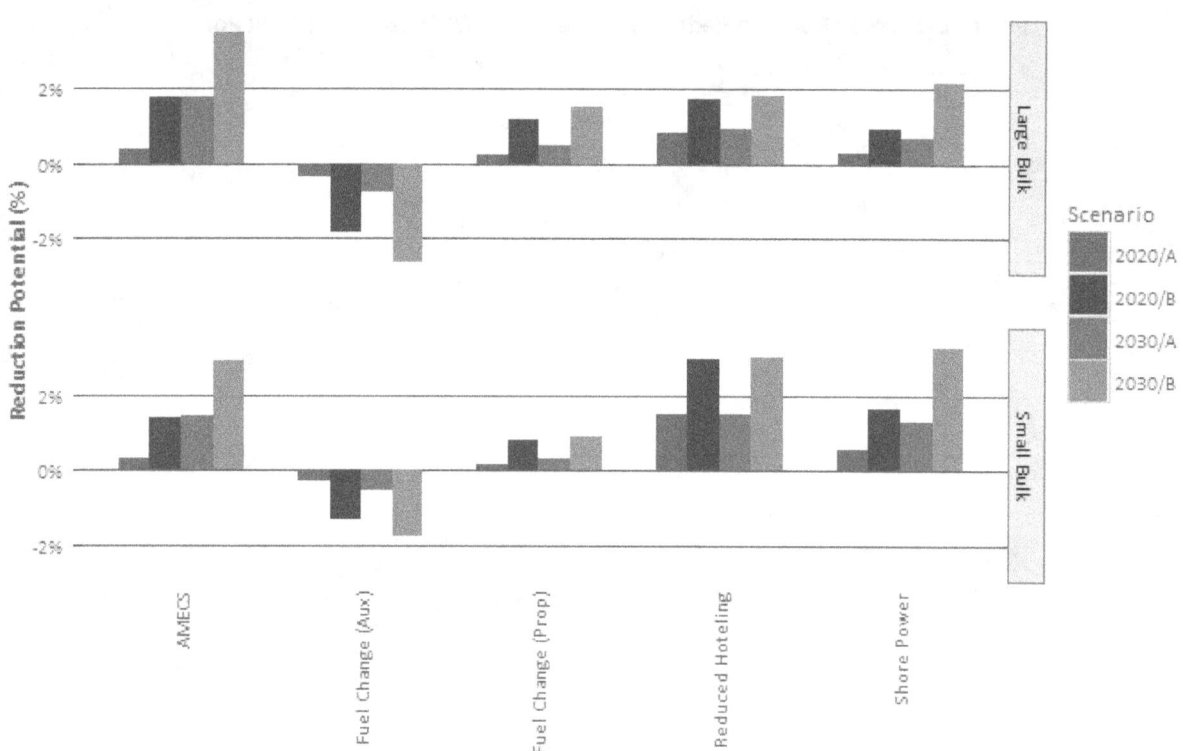

Figure D-16. BC Relative Reduction Potential of OGV Sector for Bulk Ports

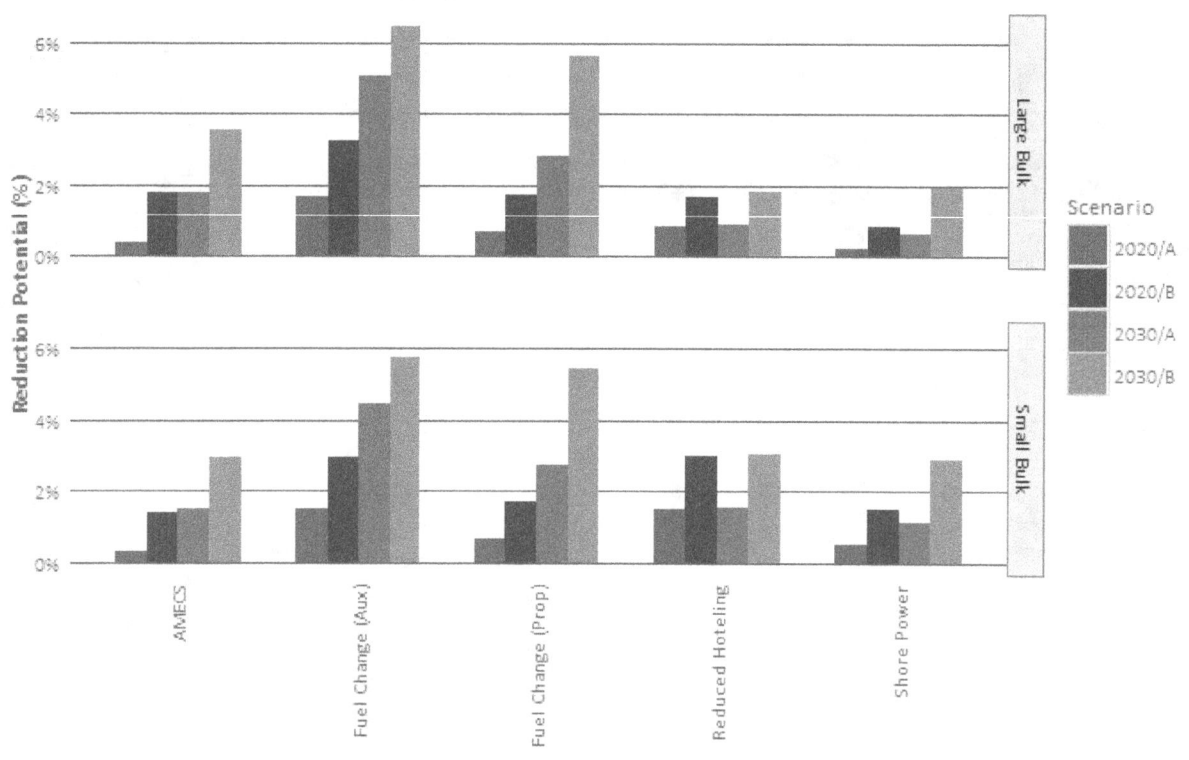

Figure D-17. SO₂ Relative Reduction Potential of OGV Sector for Bulk Ports

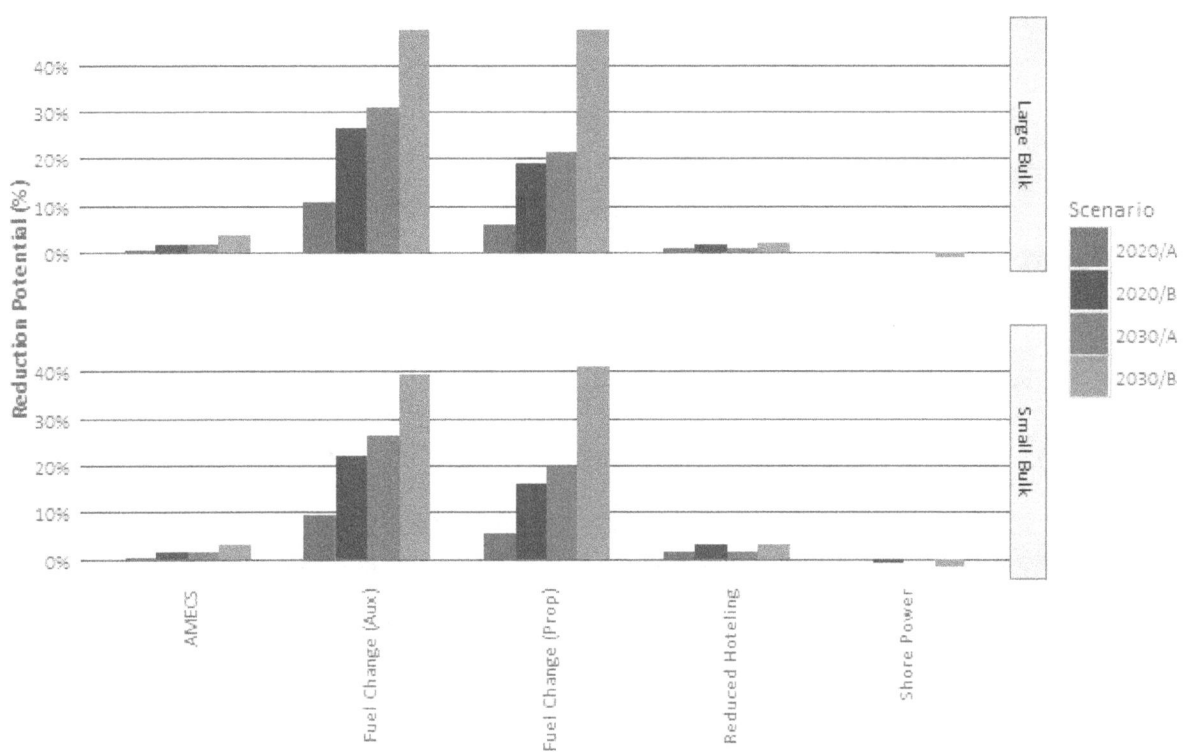

Figure D-18. CO₂ Relative Reduction Potential of OGV Sector for Bulk Ports

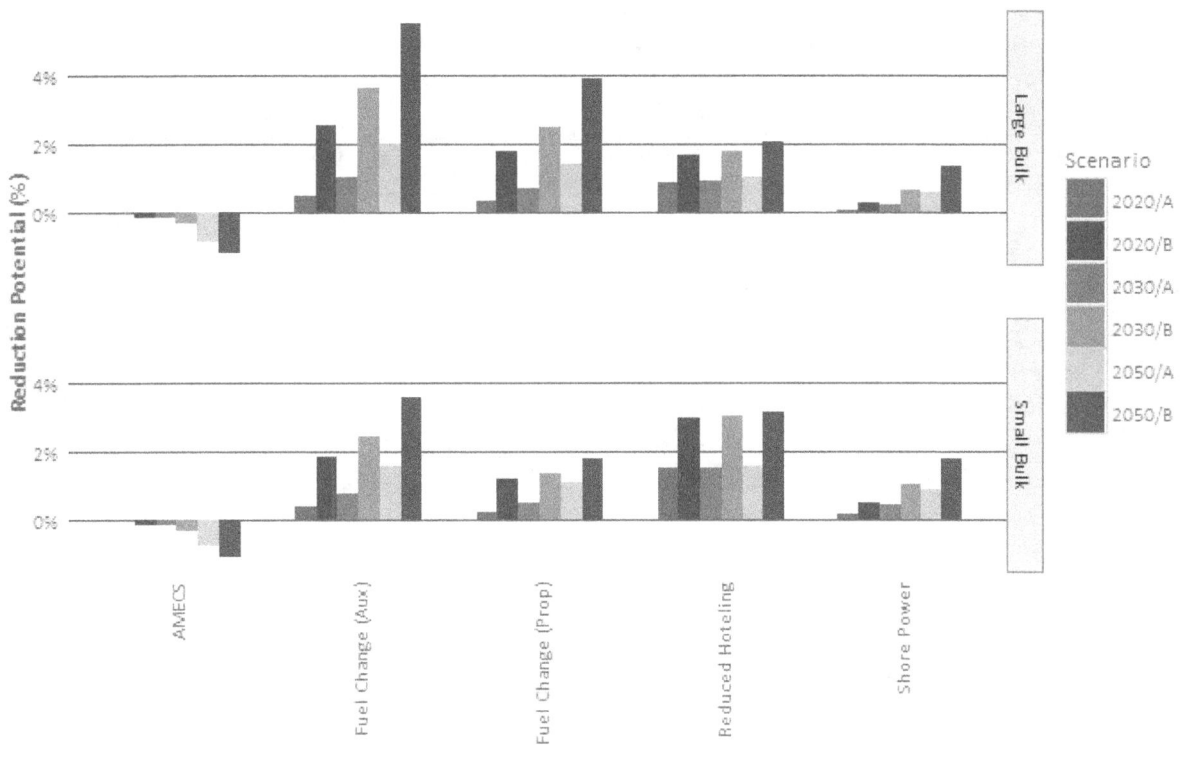

D.2.4. Passenger Ports

Figure D-19. NOx Relative Reduction Potential of OGV Sector for Passenger Ports

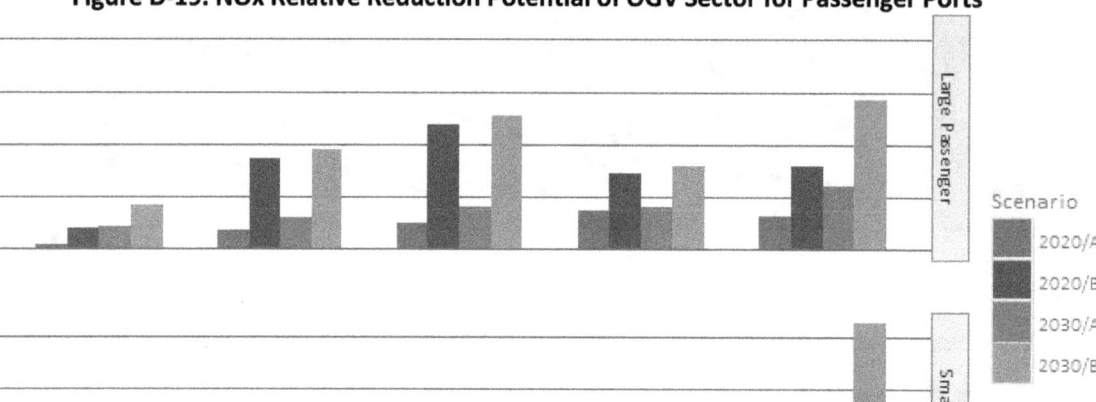

Figure D-20. PM Relative Reduction Potential of OGV Sector for Passenger Ports

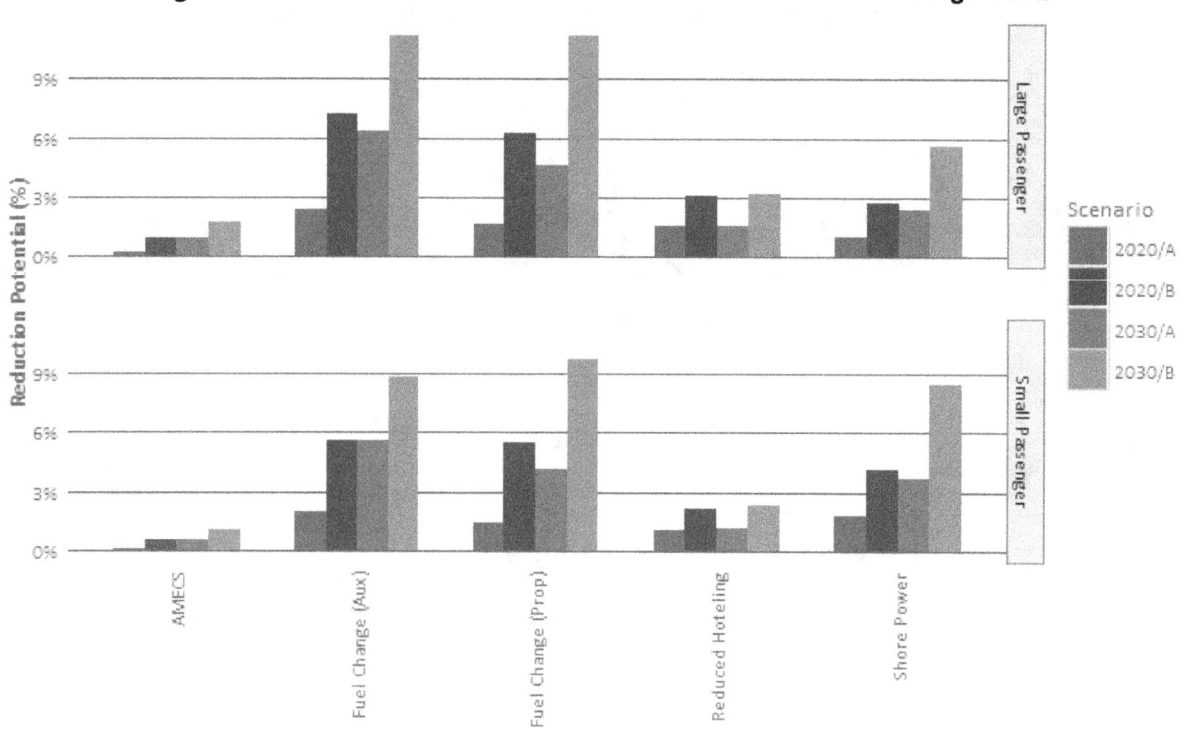

Figure D-21. HC Relative Reduction Potential of OGV Sector for Passenger Ports

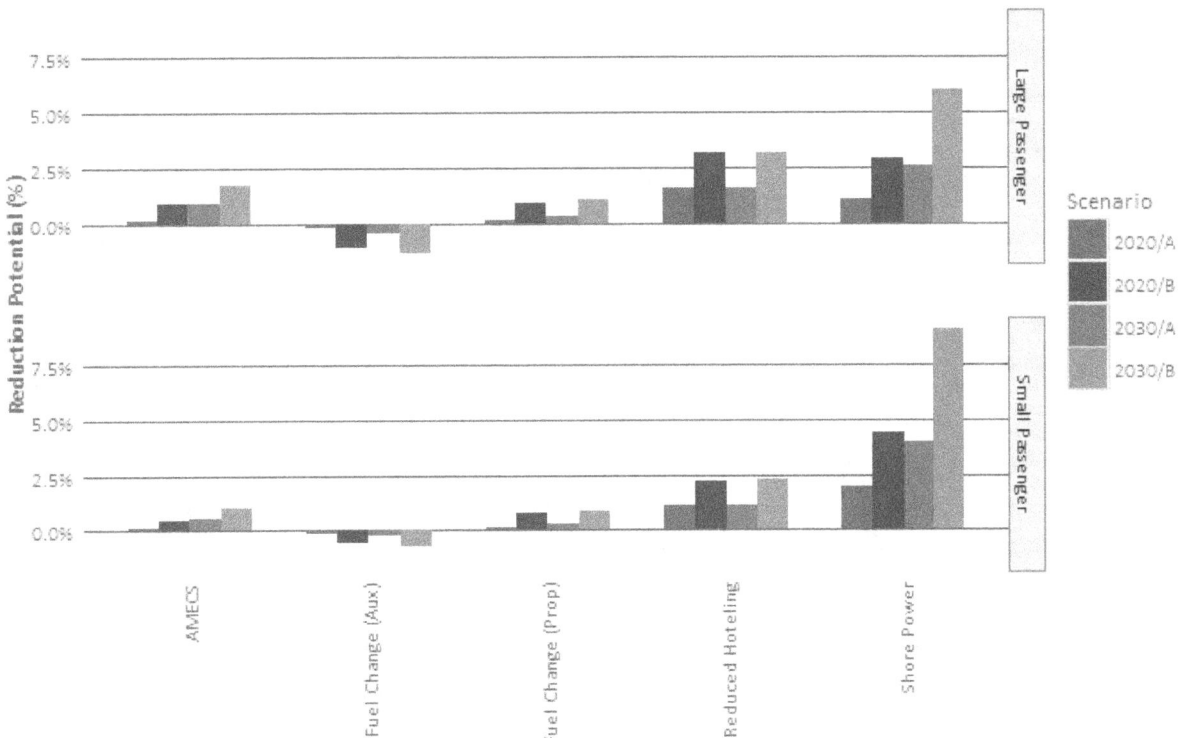

Figure D-22. BC Relative Reduction Potential of OGV Sector for Passenger Ports

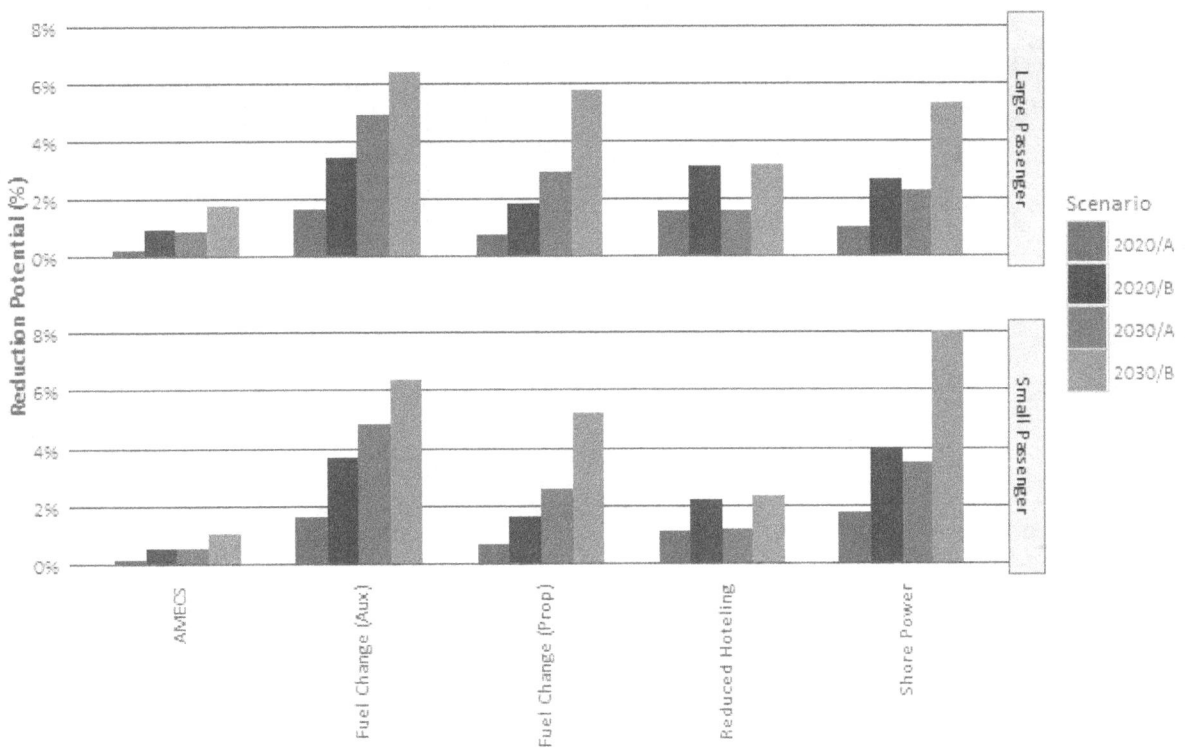

Figure D-23. SO₂ Relative Reduction Potential of OGV Sector for Passenger Ports

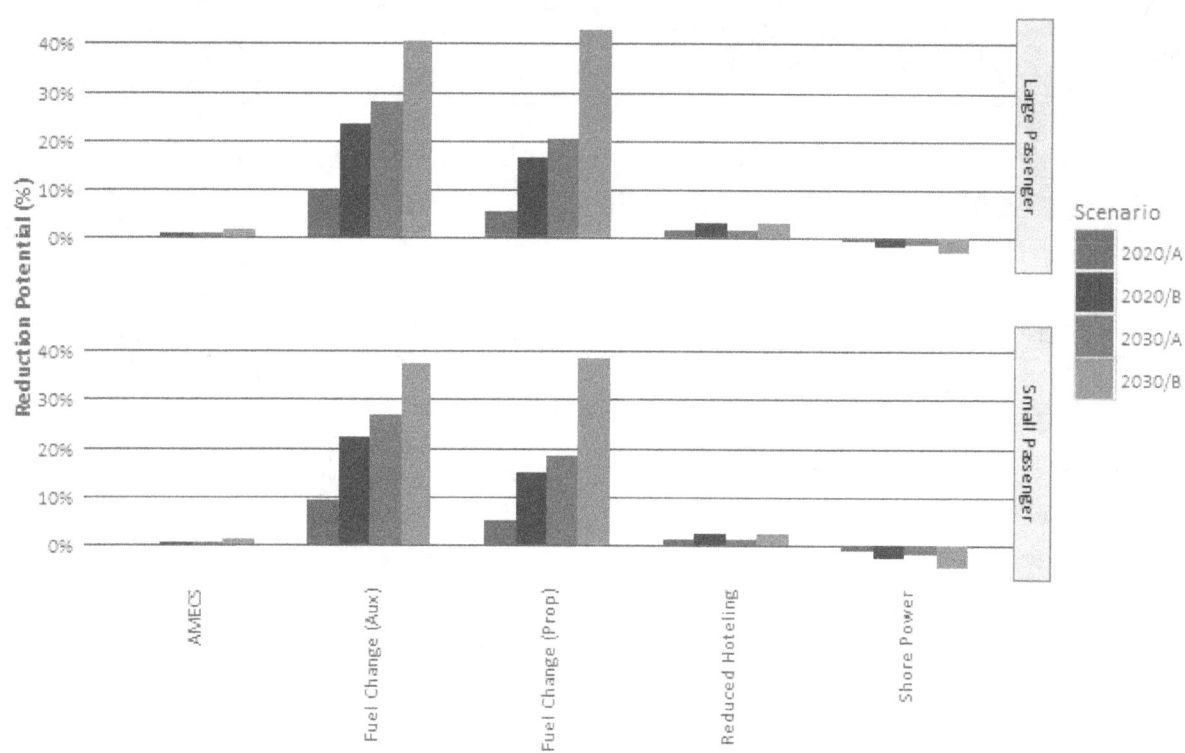

Figure D-24. CO₂ Relative Reduction Potential of OGV Sector for Passenger Ports

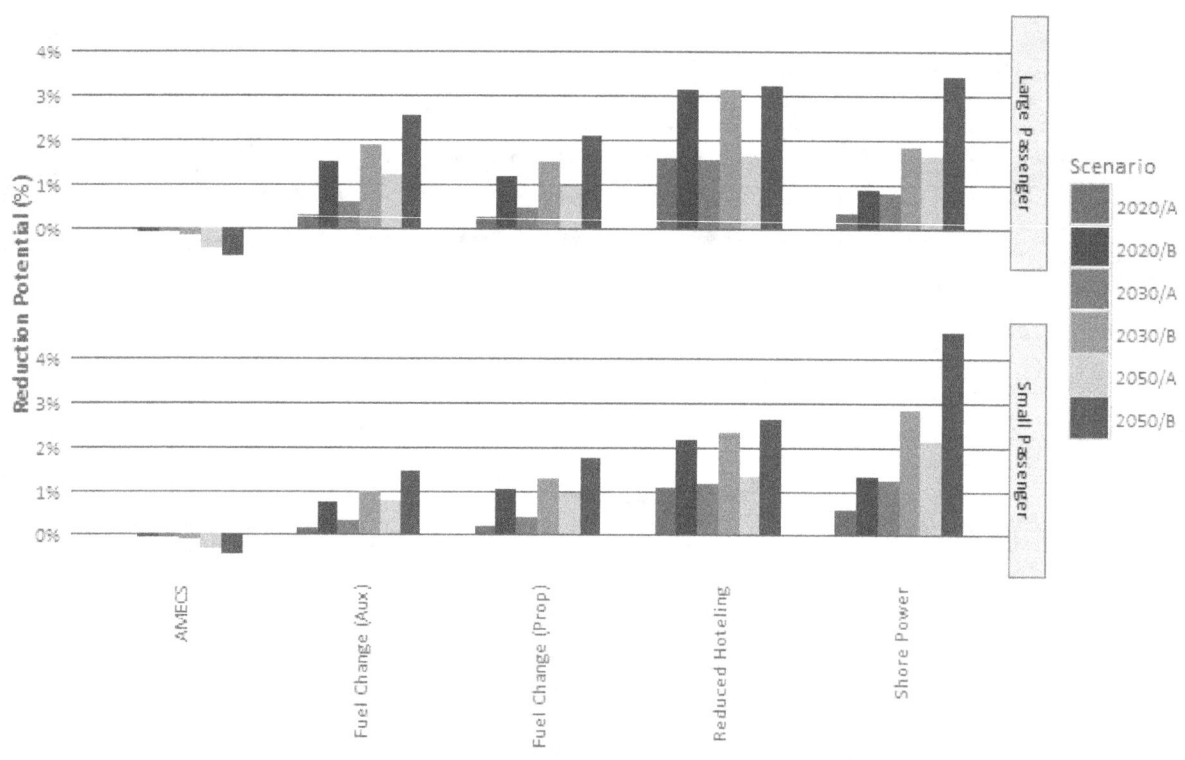

D.3. Additional Charts for Stratification Analysis of Non-OGV Sector

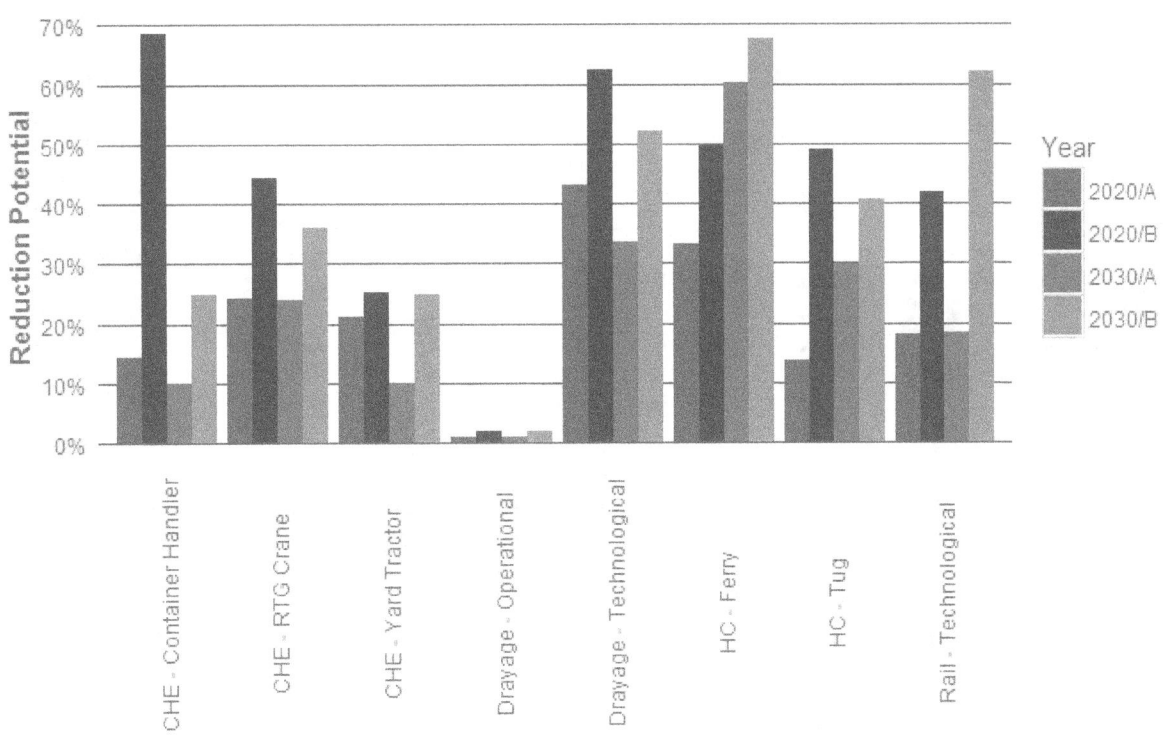

Figure D-25. NOx Relative Reduction Potential of Non-OGV Sector

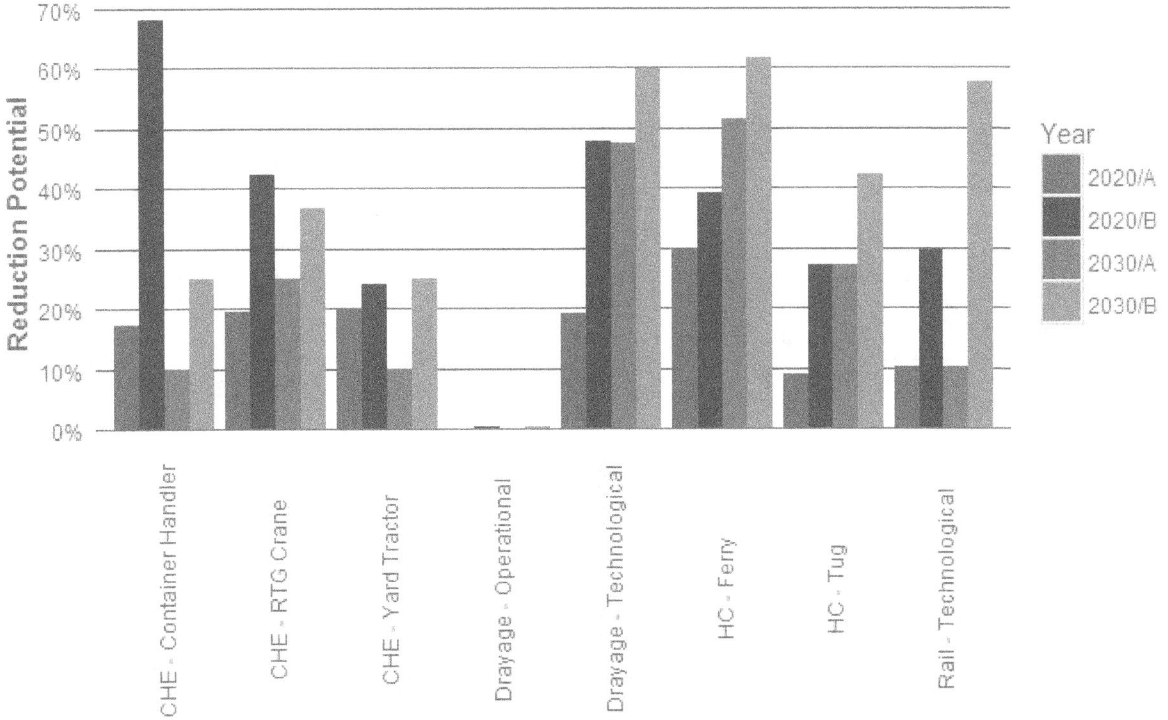

Figure D-26. PM₂.₅ Relative Reduction Potential of Non-OGV Sector

Figure D-27. Comparing BC Relative Reduction Potential of Non-OGV Sectors

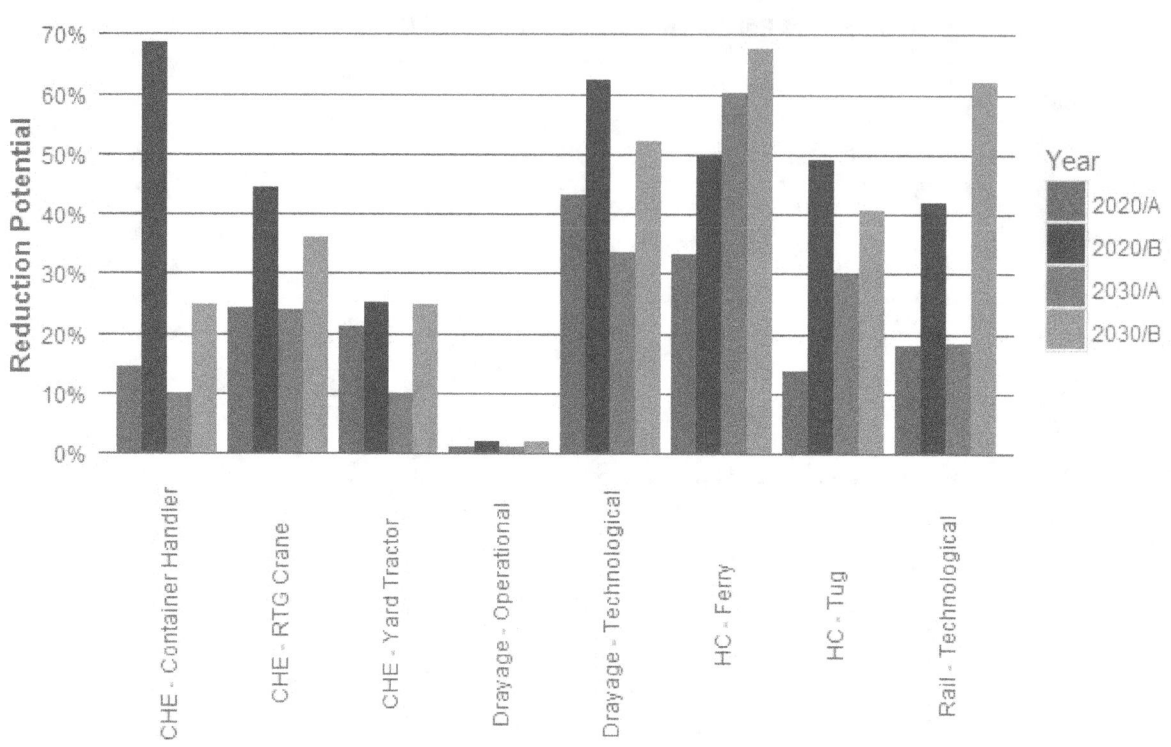

Figure D-28. Comparing CO₂ Relative Reduction Potential of Non-OGV Sectors

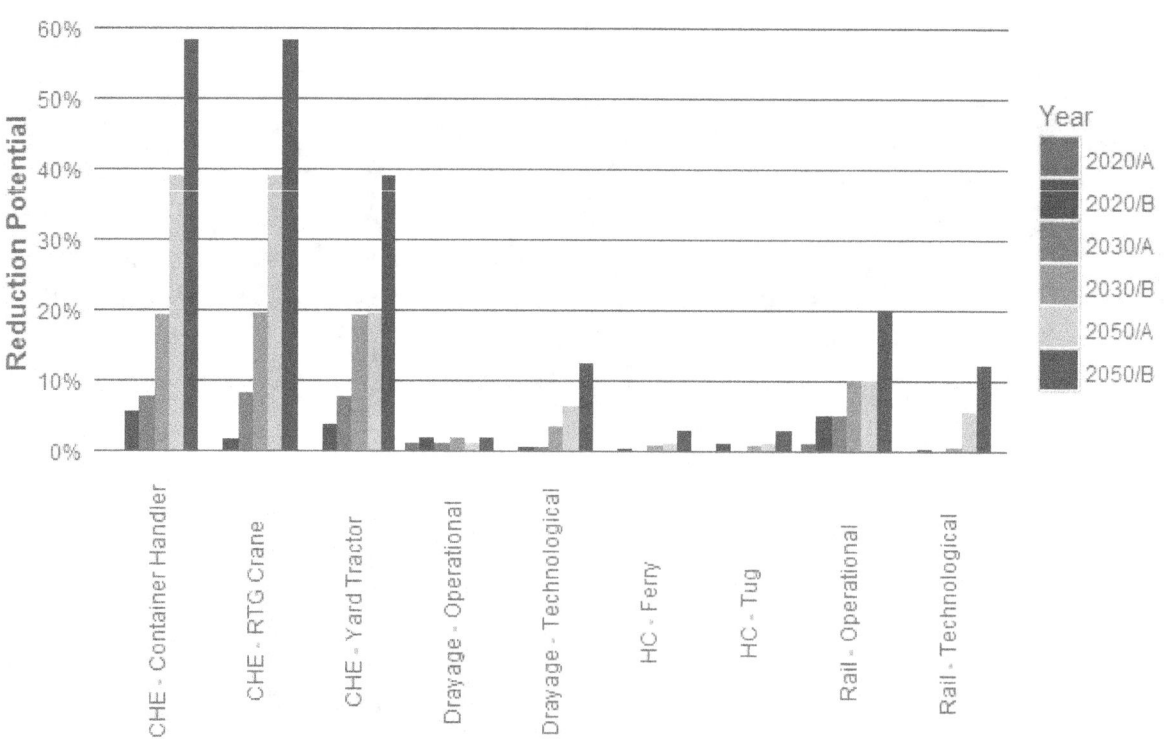

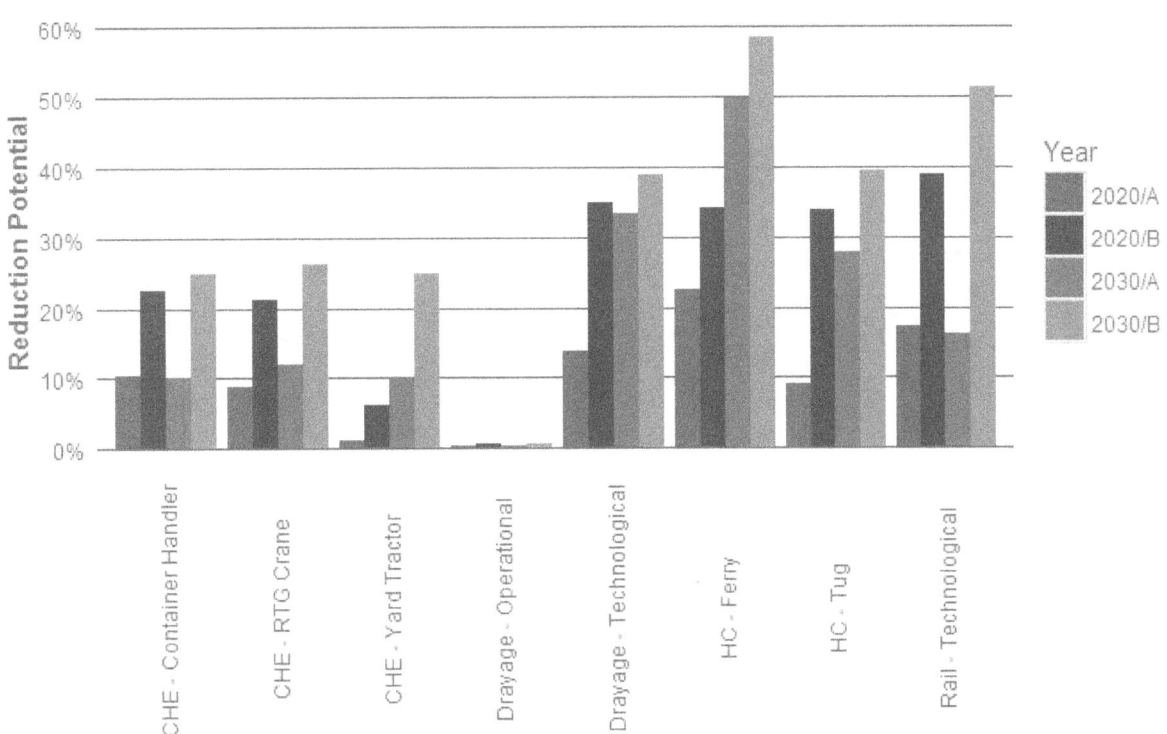

Figure D-29. Comparing VOC Relative Reduction Potential of Non-OGV Sectors

Figure D-30. Comparing Acetaldehyde Relative Reduction Potential of Non-OGV Sectors

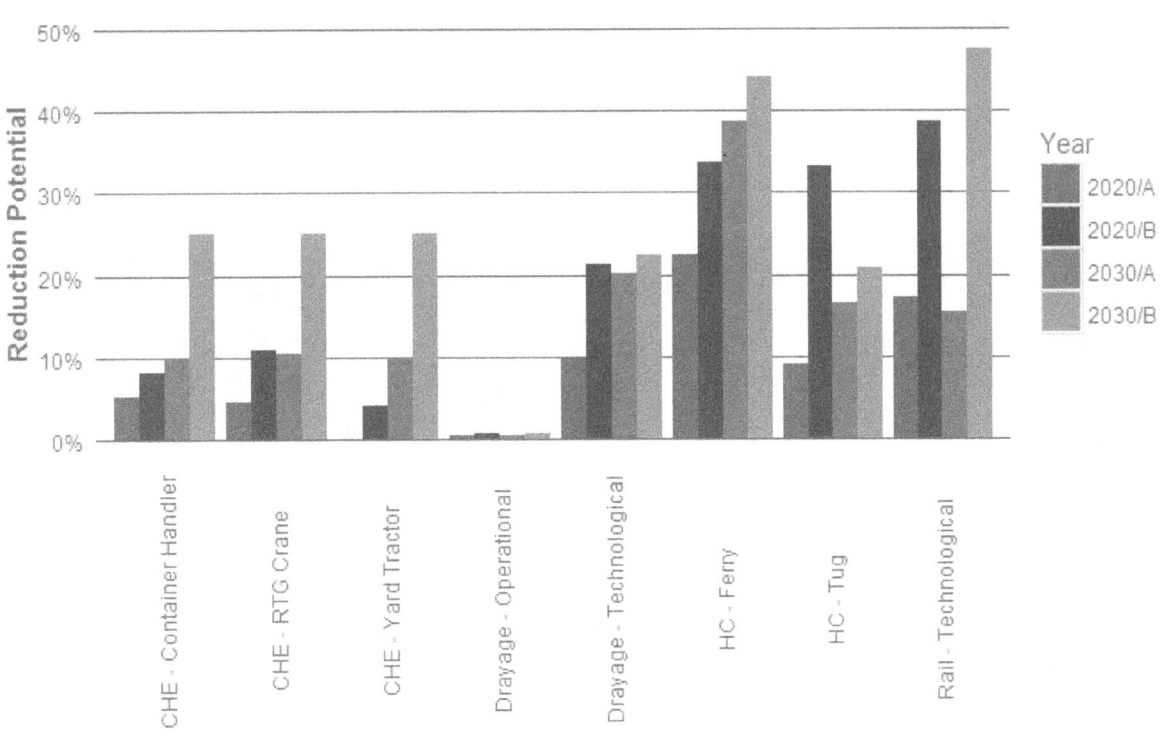

Figure D-31. Comparing Benzene Relative Reduction Potential of Non-OGV Sectors

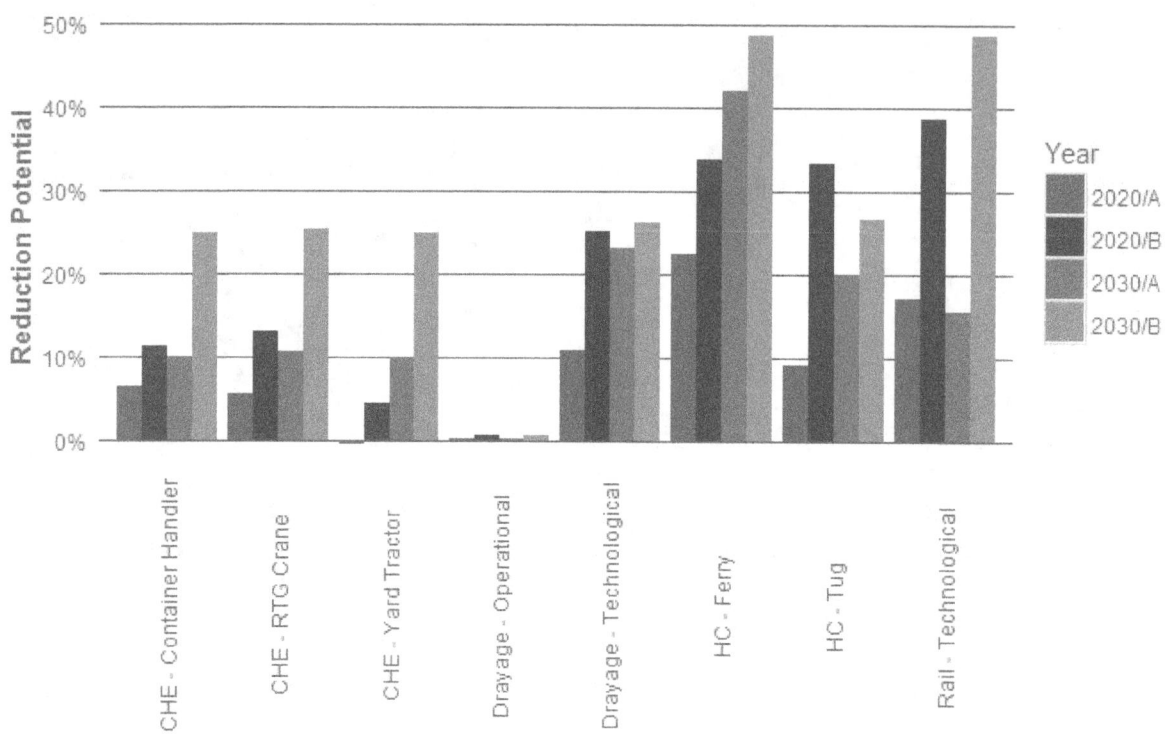

Figure D-32. Comparing Formaldehyde Relative Reduction Potential of Non-OGV Sectors

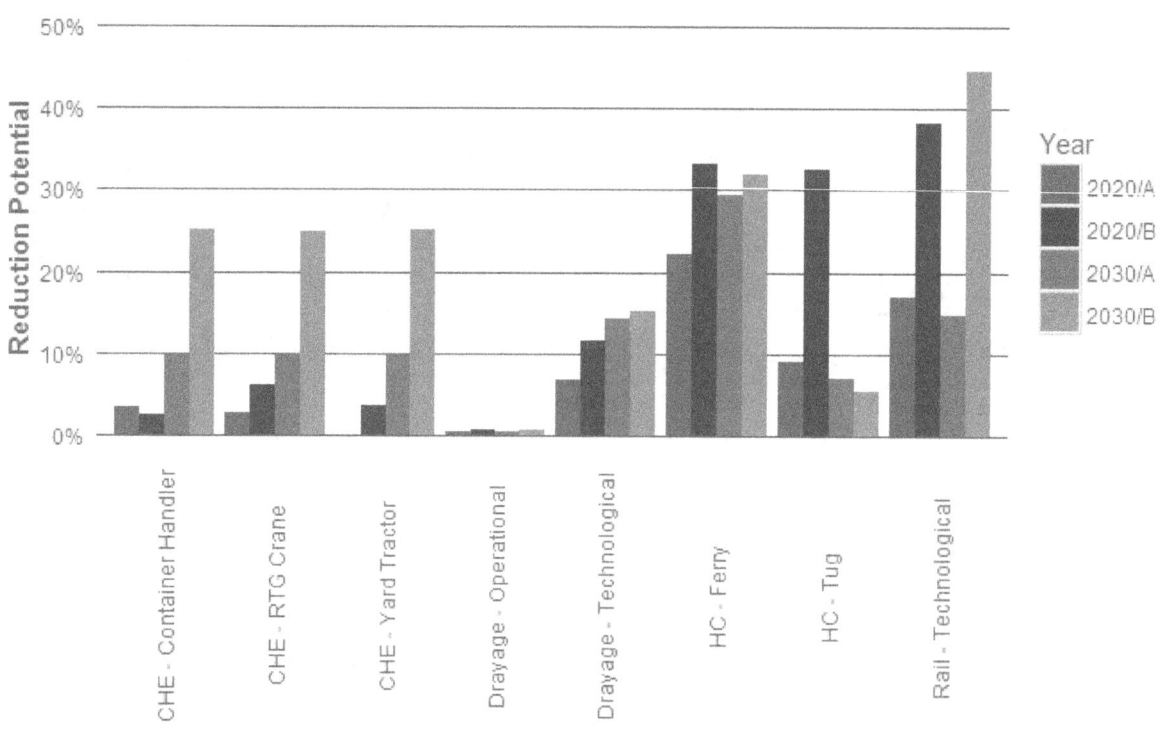